M208 Pure Mathematics

AA4

Continuity

This publication forms part of an Open University course. Details of this and other Open University courses can be obtained from the Student Registration and Enquiry Service, The Open University, PO Box 197, Milton Keynes, MK7 6BJ, United Kingdom: tel. +44 (0)870 333 4340, e-mail general-enquiries@open.ac.uk

Alternatively, you may visit the Open University website at http://www.open.ac.uk where you can learn more about the wide range of courses and packs offered at all levels by The Open University.

To purchase a selection of Open University course materials, visit the webshop at www.ouw.co.uk, or contact Open University Worldwide, Michael Young Building, Walton Hall, Milton Keynes, MK7 6AA, United Kingdom, for a brochure: tel. +44 (0)1908 858785, fax +44 (0)1908 858787, e-mail ouwenq@open.ac.uk

The Open University, Walton Hall, Milton Keynes, MK7 6AA.

First published 2006.

Edited, designed and typeset by The Open University, using the Open University TeX System.

Printed and bound in the United Kingdom by Hobbs the Printers Limited, Brunel Road, Totton, Hampshire SO40 3WX.

ISBN 0 7492 0210 6

1.1

Contents

Introduction

At the beginning of the course, we introduced techniques for sketching the graphs of many common functions, and we made various assumptions about these graphs. In this unit we justify the assumption that there are no gaps in such graphs (except 'obvious' ones at asymptotes) by introducing the concept of a *continuous function* and showing that many familiar functions are continuous. This concept is important in analysis because, in many cases, the easiest way to *prove* that a function has a property which may be intuitively obvious is to use the fact that the function is continuous.

In Section 1 we begin by revising the various fundamental operations on functions: forming *combinations*, *composites* and *inverses* of functions.

In Section 2 we define the phrase

the function f is continuous at the point a.

We give several rules which state, for example, that *combinations* of continuous functions are continuous and that *composites* of continuous functions are continuous. Using these rules, together with a list of basic continuous functions, we can deduce that many functions are continuous at each point of their domains. For example, the function

$$x \longmapsto x\sin(1/x)$$

is continuous at each point of $\mathbb{R} - \{0\}$.

We also describe the Squeeze Rule and the Glue Rule for continuous functions, which enable us to prove that certain hybrid functions are continuous. For example, we can show that the function

$$f(x) = \begin{cases} x\sin(1/x), & x \neq 0, \\ 0, & x = 0, \end{cases}$$

is continuous at 0.

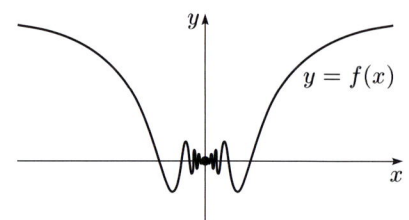

Section 3 is devoted to the *properties* of continuous functions. In particular, we prove two fundamental results called the Intermediate Value Theorem and the Extreme Value Theorem. We concentrate on the Intermediate Value Theorem and give several applications of it.

In Section 4 we discuss the Inverse Function Rule; this rule gives conditions under which a continuous function f has a continuous inverse function f^{-1}. We then use the Inverse Function Rule to show that the inverse functions of various standard functions exist and are continuous. Finally, we use the exponential function and its inverse function to define a^x, for $a > 0$ and *all* $x \in \mathbb{R}$.

In Unit AA1, Section 5, we defined a^x for $a > 0$ and $x \in \mathbb{Q}$.

— Study guide

The sections should be read in their natural order, but you are advised to spend most of your study time on Section 2 (which includes the audio) and Section 3 (which includes the video section). Section 1 is mostly revision.

1   Operations on functions

After working through this section, you should be able to:

(a) determine the domain and rule of the *sum*, *product*, *quotient* and *composite* of two real functions;

(b) determine whether a given real function has an *inverse function*.

This section revises concepts such as one-one and onto, which were introduced earlier in the course, in the context of functions whose domain and codomain are subsets of $\mathbb{R}$. Such functions can be specified in various ways. For example, the function

$$f\colon \mathbb{R} - \{0\} \longrightarrow \mathbb{R}$$
$$x \longmapsto 1/x$$

can also be written as

$$f(x) = 1/x \quad (x \in \mathbb{R} - \{0\}),$$

where the codomain of f is assumed to be $\mathbb{R}$. It can also be written as

$$f(x) = 1/x,$$

where the domain of f is assumed to be $\mathbb{R} - \{0\}$, that is, the set of values of x for which $1/x$ is defined, and where the codomain is $\mathbb{R}$. These notations illustrate the following convention.

> **Convention**   When a function f is specified just by its rule, it is to be understood that the domain of f is the set of all real numbers for which the rule is applicable and the codomain of f is $\mathbb{R}$.

We often use this convention. Sometimes, however, if we want to assert that a function has a particular property, then we may need to restrict the domain or codomain of the function. For example, the function

$$f(x) = \sin x$$

has domain and codomain $\mathbb{R}$, by our convention, but it is neither one–one nor onto. However the function

$$g\colon [-\tfrac{1}{2}\pi, \tfrac{1}{2}\pi] \longrightarrow [-1, 1]$$
$$x \longmapsto \sin x$$

has the same rule as the above function f, but g is both one-one and onto.

When we say that a function f is **defined on** a set I (usually an interval), this means that the domain of f contains the set I.

Also, recall that there are various types of interval, classified as follows.

If $a, b \in \mathbb{R}$, then:

(a, b), (a, ∞) and $(-\infty, b)$ are **open** intervals;

$[a, b]$, $[a, \infty)$ and $(-\infty, b]$ are **closed** intervals;

$[a, b)$ and $(a, b]$ are **half-open** intervals;

$\mathbb{R} = (-\infty, \infty)$ is both open and closed.

See Unit I2, Section 2.

See Subsection 4.2.

For example, the function $f(x) = 1/x$ is defined on $[1, 2]$, but not on $[-1, 1]$.

See Unit I1, Subsection 1.1.

Some texts use the notation $]a, b[$ for open intervals.

1.1 Sums, products and quotients of functions

Let f and g be the functions

$$f(x) = 1/x \quad (x \in \mathbb{R} - \{0\}) \quad \text{and} \quad g(x) = \sin x \quad (x \in \mathbb{R}).$$

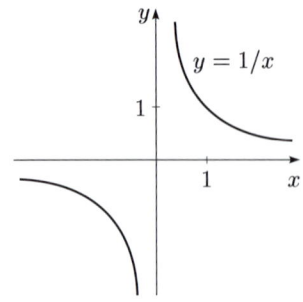
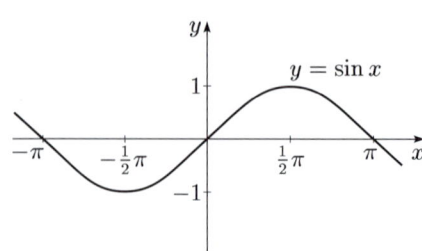

We use the expressions $f + g$, fg and f/g to denote the functions

$$(f + g)(x) = f(x) + g(x) = 1/x + \sin x \quad (x \in \mathbb{R} - \{0\}),$$

$$(fg)(x) = f(x)g(x) = \frac{\sin x}{x} \quad (x \in \mathbb{R} - \{0\})$$

and

$$\left(\frac{f}{g}\right)(x) = \frac{f(x)}{g(x)} = \frac{1}{x \sin x} \quad (x \in \mathbb{R} - \{n\pi : n \in \mathbb{Z}\}).$$

The set $\{x : \sin x = 0\}$ is $\{n\pi : n \in \mathbb{Z}\}$.

The domains of $f + g$, fg and f/g include only those points for which f and g are *both* defined. Also, when forming the quotient f/g, we must exclude from the domain all the points x such that $g(x) = 0$.

We often use the word 'point' to mean 'number'.

Definitions   If $f \colon A \longrightarrow \mathbb{R}$ and $g \colon B \longrightarrow \mathbb{R}$, then the **sum** $f + g$ is the function with domain $A \cap B$ and rule

$$(f + g)(x) = f(x) + g(x);$$

the **multiple** λf is the function with domain A and rule

$$(\lambda f)(x) = \lambda f(x), \quad \text{for } \lambda \in \mathbb{R};$$

the **product** fg is the function with domain $A \cap B$ and rule

$$(fg)(x) = f(x)g(x);$$

the **quotient** f/g is the function with domain $A \cap B - \{x : g(x) = 0\}$ and rule

$$(f/g)(x) = f(x)/g(x).$$

Remark   Often, we wish to form the sum, product or quotient of functions f and g which have the *same* domain, A say. In this case, A is also the domain of $f + g$ and fg, and the domain of f/g is $A - \{x : g(x) = 0\}$.

Exercise 1.1   Let f and g be the functions

$$f(x) = e^x \quad (x \in \mathbb{R}) \quad \text{and} \quad g(x) = \tan x \quad (x \in (-\tfrac{1}{2}\pi, \tfrac{1}{2}\pi)).$$

Determine the domain and rule of the functions $f + g$, fg and f/g.

— 1.2 Composite functions

Let f and g be functions. Then the composite function $g \circ f$ is the function defined by the rule

$$(g \circ f)(x) = g(f(x)),$$

where we apply first f and then g. Again, we must exclude from the domain all the points x which lead to an expression that is not defined.

For example, if

$$f(x) = 1/x \quad (x \in \mathbb{R} - \{0\}) \quad \text{and} \quad g(x) = \sin x \quad (x \in \mathbb{R}),$$

then $g \circ f$ is the function

$$(g \circ f)(x) = \sin \frac{1}{x} \quad (x \in \mathbb{R} - \{0\}),$$

whereas $f \circ g$ is the function

$$(f \circ g)(x) = \frac{1}{\sin x} \quad (x \in \mathbb{R} - \{n\pi : n \in \mathbb{Z}\}).$$

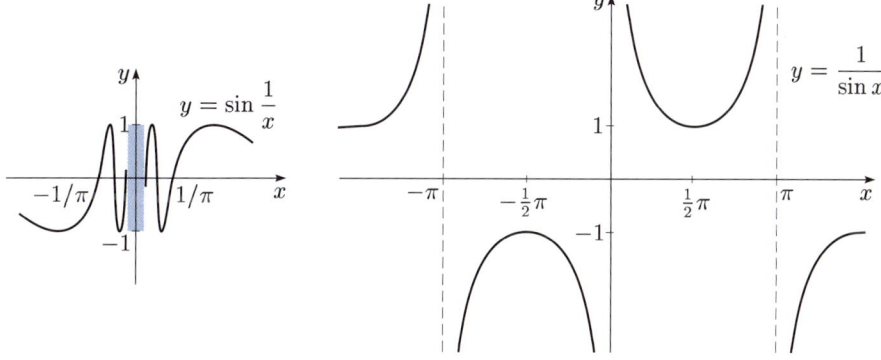

More generally, if $f \colon A \longrightarrow \mathbb{R}$ and $g \colon B \longrightarrow \mathbb{R}$, then $g(f(x))$ is defined if and only if x lies in the domain of f and $f(x)$ lies in the domain of g.

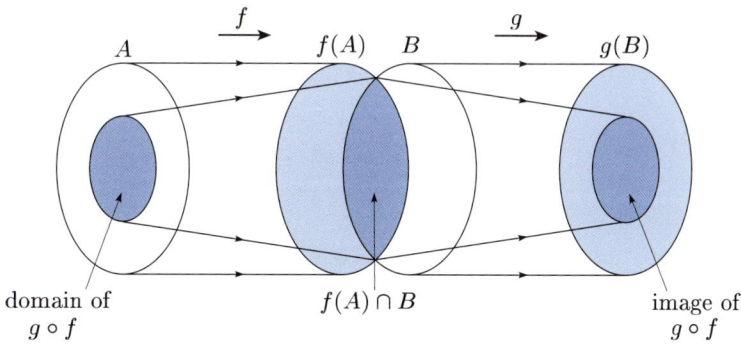

Definition   If $f \colon A \longrightarrow \mathbb{R}$ and $g \colon B \longrightarrow \mathbb{R}$, then the **composite** $g \circ f$ has domain

$$\{x \in A : f(x) \in B\}$$

and rule

$$(g \circ f)(x) = g(f(x)).$$

Remark   This definition allows us to form the composite of *any* two functions, though in some cases we may obtain a strange function whose domain is the empty set $\varnothing$. For example, if

$$f(x) = -x^2 - 1 \quad (x \in \mathbb{R}) \quad \text{and} \quad g(x) = \sqrt{x} \quad (x \in [0, \infty)),$$

then the composite function $g \circ f$ is defined nowhere and so has domain $\varnothing$.

The rule of $g \circ f$ is

$$g(f(x)) = \sqrt{-x^2 - 1}.$$

Frequently, however, it happens that the image

$$f(A) = \{f(x) : x \in A\} \subseteq B.$$

In this case, the set A is also the domain of $g \circ f$.

The image of a function was introduced in Unit I2, Section 2.

Exercise 1.2   Let f and g be the functions

$$f(x) = \sqrt{x} \quad (x \in [0, \infty)) \quad \text{and} \quad g(x) = \sin x \quad (x \in \mathbb{R}).$$

Determine the domain and rule of the composites $f \circ g$ and $g \circ f$.

1.3 Inverse functions

Let f be the function

$$f(x) = 2x \quad (x \in \mathbb{R}).$$

Then, for each number y in $\mathbb{R}$, there is a unique number $x = \frac{1}{2}y$ in the domain of f such that

$$f(x) = f(\tfrac{1}{2}y) = 2 \times \tfrac{1}{2}y = y.$$

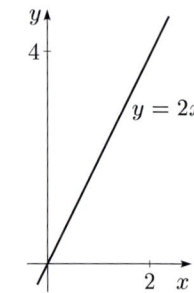

The corresponding function $g(y) = \frac{1}{2}y$ is called the *inverse function* of f because it undoes the effect of f; that is,

$$g(f(x)) = x, \quad \text{for } x \in \mathbb{R},$$

and

$$f(g(y)) = y, \quad \text{for } y \in \mathbb{R}.$$

However, not every function has an inverse function. For example, consider the function

$$f(x) = x^2 \quad (x \in \mathbb{R}).$$

Since $f(2) = 4 = f(-2)$, we cannot assign a unique value x in the domain of f such that $f(x) = 4$. The problem here is that this function f is not one-one. In general, it is possible to define the inverse function of a function only if that function is one-one.

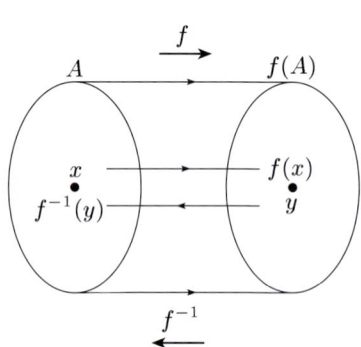

> **Definition**   Let $f : A \longrightarrow \mathbb{R}$ be a one-one function. Then the **inverse function** f^{-1} of f has domain $f(A)$ and rule
>
> $$f^{-1}(y) = x, \quad \text{where } y = f(x).$$

For some functions f, we can find the inverse function f^{-1} directly by solving the equation $y = f(x)$ algebraically to obtain x in terms of y.

Example 1.1   Prove that the following function has an inverse function, and find the domain and rule of this inverse function:

$$f(x) = \frac{1}{1-x} \quad (x \in (-\infty, 1)).$$

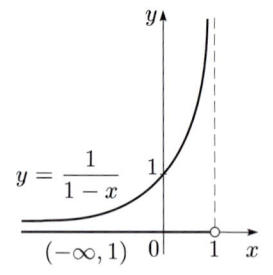

Solution   First we solve the equation

$$y = \frac{1}{1-x}$$

to give x in terms of y. Rearranging this equation, we obtain

$$x = 1 - \frac{1}{y}.$$

This shows that f is one-one, so f has an inverse function with rule $f^{-1}(y) = 1 - 1/y$.

Now we find the image of f. For each x in the domain $(-\infty, 1)$, we have $x < 1$, so

$$f(x) = \frac{1}{1-x} > 0.$$

Thus $f((-\infty, 1)) \subseteq (0, \infty)$.

On the other hand, for each y in $(0, \infty)$, we have $1/y > 0$, so

$$x = 1 - \frac{1}{y} \in (-\infty, 1).$$

Thus $f((-\infty, 1)) \supseteq (0, \infty)$, so $f((-\infty, 1)) = (0, \infty)$.

Hence the domain of f^{-1} is $(0, \infty)$, so

$$f^{-1}(y) = 1 - \frac{1}{y} \quad (y \in (0, \infty)). \quad \blacksquare$$

We have

$$y = \frac{1}{1-x} \Leftrightarrow \frac{1}{y} = 1 - x$$

$$\Leftrightarrow x = 1 - \frac{1}{y}.$$

Remarks

1. Usually, when defining a function, we write x for the domain variable. To conform with this practice, we can rewrite the inverse function f^{-1} in Example 1.1 as

    $$f^{-1}(x) = 1 - \frac{1}{x} \quad (x \in (0, \infty)).$$

2. The graph $y = f^{-1}(x)$ is obtained by reflecting the graph $y = f(x)$ in the line $y = x$.

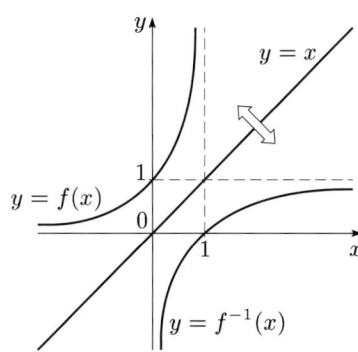

Proving that a function is one-one

We have seen that if $f \colon A \longrightarrow \mathbb{R}$ is one-one, then f has an inverse function f^{-1} with domain $f(A)$. For the function f considered in Example 1.1, it is possible to determine f^{-1} explicitly by solving the equation $y = f(x)$ to obtain x in terms of y. Unfortunately, it is often impossible to solve the equation $y = f(x)$ in this way. Nevertheless, it may still be possible to prove that f has an inverse function f^{-1} by showing that f is one-one.

40

Earlier in the course, we showed that various functions f are one-one by proving algebraically that

See Unit I2, Subsection 2.3.

if $f(x_1) = f(x_2)$, then $x_1 = x_2$.

However, this algebraic method can be used only for fairly simple functions.

Another way of showing that f is one-one is by proving that f is *strictly increasing* or *strictly decreasing*.

> **Definition**   A function f defined on an interval I is
>
> **increasing** on I if
> $$x_1 < x_2 \;\Rightarrow\; f(x_1) \le f(x_2), \quad \text{for } x_1, x_2 \in I;$$
> **strictly increasing** on I if
> $$x_1 < x_2 \;\Rightarrow\; f(x_1) < f(x_2), \quad \text{for } x_1, x_2 \in I;$$
> **decreasing** on I if
> $$x_1 < x_2 \;\Rightarrow\; f(x_1) \ge f(x_2), \quad \text{for } x_1, x_2 \in I;$$
> **strictly decreasing** on I if
> $$x_1 < x_2 \;\Rightarrow\; f(x_1) > f(x_2), \quad \text{for } x_1, x_2 \in I;$$
> **monotonic** on I if f is either increasing on I or decreasing on I;
>
> **strictly monotonic** on I if f is either strictly increasing on I or strictly decreasing on I.

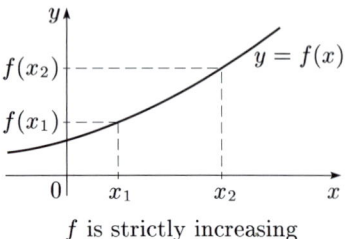

f is strictly increasing

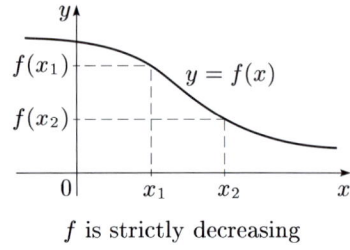

f is strictly decreasing

Remark   If the interval I is the domain of f, then we omit 'on I' and say, for example,

f is increasing.

The most powerful technique for proving that a function f is increasing or decreasing is to compute the derivative f' of f and examine the sign of $f'(x)$. Here, however, we consider only those functions which can be proved to be increasing or decreasing by manipulating inequalities, rather than by using calculus.

We used this technique in Unit I1 and we shall use it again in Analysis Block B.

For example, if $n \in \mathbb{N}$, then the function

$$f(x) = x^n \quad (x \in [0, \infty))$$

is strictly increasing; and if n is odd, then the function

This follows from Unit AA1, Subsection 2.1, Rule 5.

$$f(x) = x^n \quad (x \in \mathbb{R})$$

is strictly increasing.

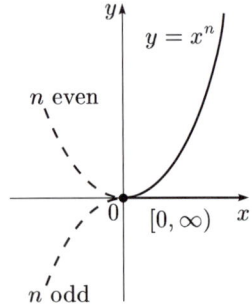

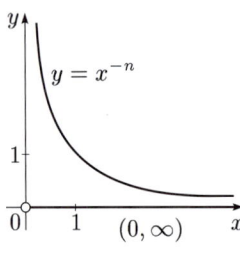

Similarly, if $n \in \mathbb{N}$, then the function

$$f(x) = x^{-n} \quad (x \in (0, \infty))$$

is strictly decreasing.

Any function that is strictly monotonic must be one-one. Here is an example of how this property can be used.

Example 1.2   Prove that the function

$$f(x) = x^5 + x - 1 \quad (x \in \mathbb{R})$$

is one-one.

Solution   If $x_1 < x_2$, then $x_1^5 < x_2^5$, so

$$x_1^5 + x_1 - 1 < x_2^5 + x_2 - 1; \quad \text{that is,} \quad f(x_1) < f(x_2).$$

Hence f is strictly increasing and thus one-one.   ■

Exercise 1.3   Prove that the following functions are one-one.
(a) $f(x) = x^4 + 2x + 3 \quad (x \in [0, \infty))$
(b) $f(x) = x^2 - 1/x \quad (x \in (0, \infty))$

Determining the image

If the function $f: A \longrightarrow \mathbb{R}$ is strictly increasing or strictly decreasing, then f is one-one, so f has an inverse function f^{-1} with domain $f(A)$. However, it is not always easy to determine the image $f(A)$.

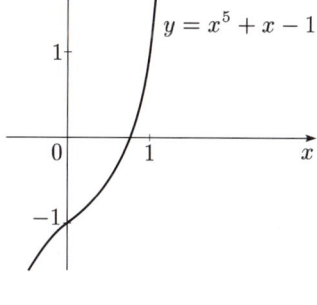

For example, consider the function

$$f(x) = x^5 + x - 1 \quad (x \in \mathbb{R}).$$

We saw in Example 1.2 that f is one-one, so f has an inverse function with domain $f(\mathbb{R})$. Since f is strictly increasing, it seems likely that $f: \mathbb{R} \longrightarrow \mathbb{R}$ is onto, so $f(\mathbb{R}) = \mathbb{R}$, and we have sketched the graph $y = f(x)$ as though this is the case. But how can we *prove* that $f(\mathbb{R}) = \mathbb{R}$? To do this we want to show that, for each $y \in \mathbb{R}$, there is an x such that

$$f(x) = x^5 + x - 1 = y.$$

Unfortunately, we cannot find such an x by solving this equation algebraically to obtain x in terms of y. Could it be that the graph $y = x^5 + x - 1$ actually has some 'gaps' or 'jumps' in it? We would be very surprised if gaps do occur, but how can we prove that they do not?

To answer this question, we need the concept of *continuity*, which we introduce in Section 2.

Further exercises

Exercise 1.4   Let f and g be the functions

$$f(x) = \tan x \quad (x \in (-\tfrac{1}{2}\pi, \tfrac{1}{2}\pi)) \quad \text{and} \quad g(x) = \sqrt{x} \quad (x \in [0, \infty)).$$

Determine the domain and the rule of the following functions.
(a) $f + g$ (b) fg (c) f/g (d) $f \circ g$ (e) $g \circ f$

Exercise 1.5   Prove that the following function has an inverse function, and find the domain and rule of this inverse function.

$$f(x) = \frac{x - 1}{x + 2} \quad (x \in (-2, \infty))$$

In Exercise 1.5 it helps to write

$$\frac{x - 1}{x + 2} = 1 - \frac{3}{x + 2}.$$

Exercise 1.6   By checking the definition, prove that each of the following functions is strictly monotonic.

(a) $f(x) = x^3 + 1 - 1/x^2 \quad (x \in (0, \infty))$

(b) $f(x) = \dfrac{1}{(1 + x^3)^2} \quad (x \in [0, \infty))$

2   Continuous functions

After working through this section, you should be able to:

(a) explain the meaning of the phrase 'f is *continuous* at a';

(b) use the Combination Rules, the Composition Rule, the Squeeze Rule and the Glue Rule for continuous functions;

(c) recognise certain *basic* continuous functions.

2.1 What is continuity?

At the end of Section 1 we asked whether the graph of the function

$$f(x) = x^5 + x - 1 \quad (x \in \mathbb{R})$$

has any gaps. Put another way, can we draw the graph $y = x^5 + x - 1$ without lifting a pen from the paper? In this section we show that this graph cannot have any gaps because the function f is *continuous*.

The first object of the audio is to *define* the phrase

    f is continuous at the point a.

To accord with our intuitive ideas, we wish to define this concept in such a way that the following two functions are continuous at the point a.

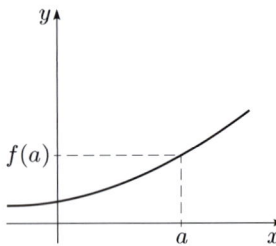
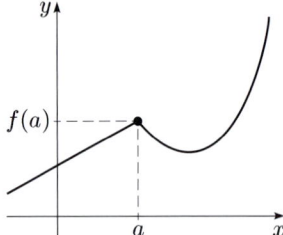

On the other hand, we wish to formulate our definition so that the following two functions are *not* continuous at the point a.

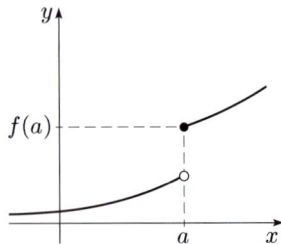
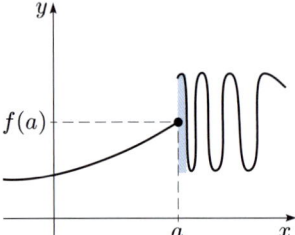

Our definition must say, in a precise way, that

    if x tends to a, then $f(x)$ tends to $f(a)$.

There are several ways of making this idea precise. Here we adopt a definition that involves the *convergence of sequences*, as this enables us to use known results about sequences to prove that many common functions are continuous.

In Analysis Block B we give another definition of continuity which is convenient for dealing with more unusual functions.

In the audio frames we use two inequalities proved earlier:

$$|a - b| \geq \big||a| - |b|\big|, \quad \text{for } a, b \in \mathbb{R},$$

which is the backwards form of the Triangle Inequality, and

See Unit AA1, Subsection 3.1.

$$\sqrt{|a - b|} \geq |\sqrt{a} - \sqrt{b}|, \quad \text{for } a, b \geq 0.$$

We also use two further inequalities

See Unit AA1, Subsection 3.3, Frame 3.

$$|\sin x| \leq |x|, \quad \text{for } x \in \mathbb{R}, \quad \text{and} \quad 1 + x \leq e^x \leq \frac{1}{1 - x}, \quad \text{for } |x| < 1.$$

We prove these inequalities in Subsection 2.3.

Before starting the audio, try the following exercise. Its solution is discussed in Frames 1 and 2. Here the sequences are denoted by $\{x_n\}$ rather than $\{a_n\}$ because, as you will see, they represent points on the x-axis.

— **Exercise 2.1** Let $\{x_n\}$ be a sequence such that $\lim\limits_{n \to \infty} x_n = 2$. Determine the limits of the following sequences.

(a) $\{3x_n\}$ (b) $\{x_n^2\}$ (c) $\{1/x_n\}$

Listen to the audio as you work through the frames.

Audio

3. Continuity: intuitive idea

Cloud bubble: $f: A \to \mathbb{R}$ $a \in A$

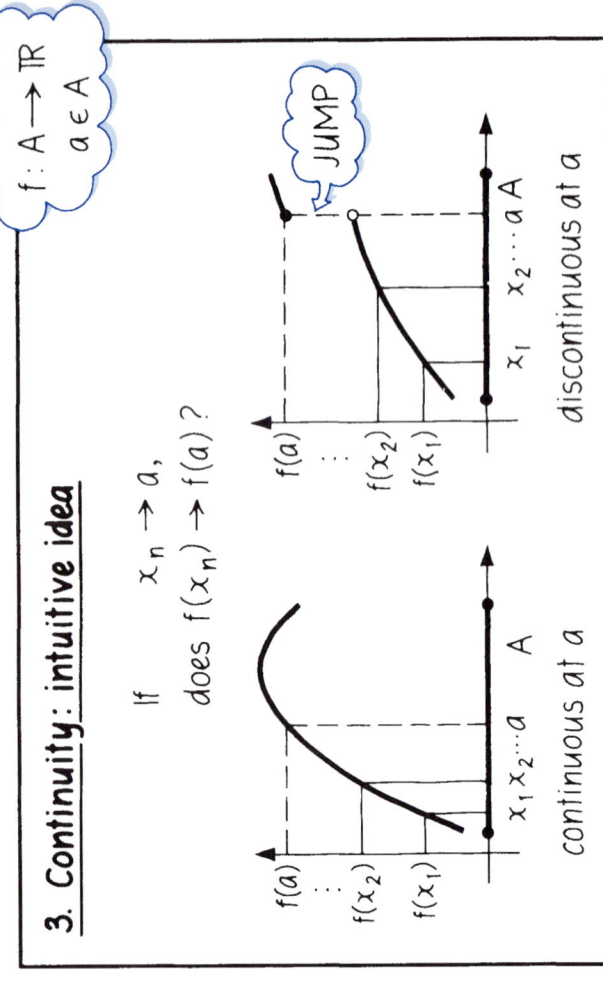

If $x_n \to a$,

does $f(x_n) \to f(a)$?

JUMP

continuous at a

discontinuous at a

4. Continuity: definition

Cloud bubble: $f: A \to \mathbb{R}$ $a \in A$

- f is **continuous at a** if:

  for each sequence $\{x_n\}$ in A such that

  $x_n \to a$, we have $f(x_n) \to f(a)$.

  i.e. $\boxed{x_n \to a} \Rightarrow \boxed{f(x_n) \to f(a)}$

- f is **continuous** (on A) means

  f is continuous at each $a \in A$.

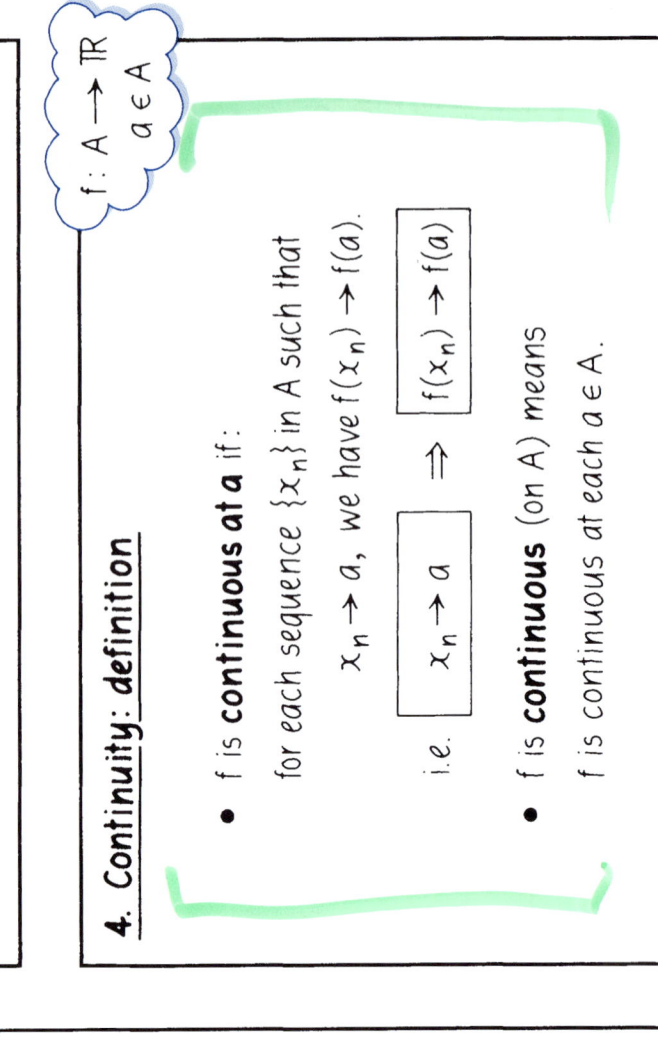

1. Exercise 2.1(a) revisited

Rewrite

$\boxed{x_n \to 2} \Rightarrow \boxed{3x_n \to 6}$

as

$\boxed{x_n \to 2} \Rightarrow \boxed{f(x_n) \to f(2)}$

where $f(x) = 3x$

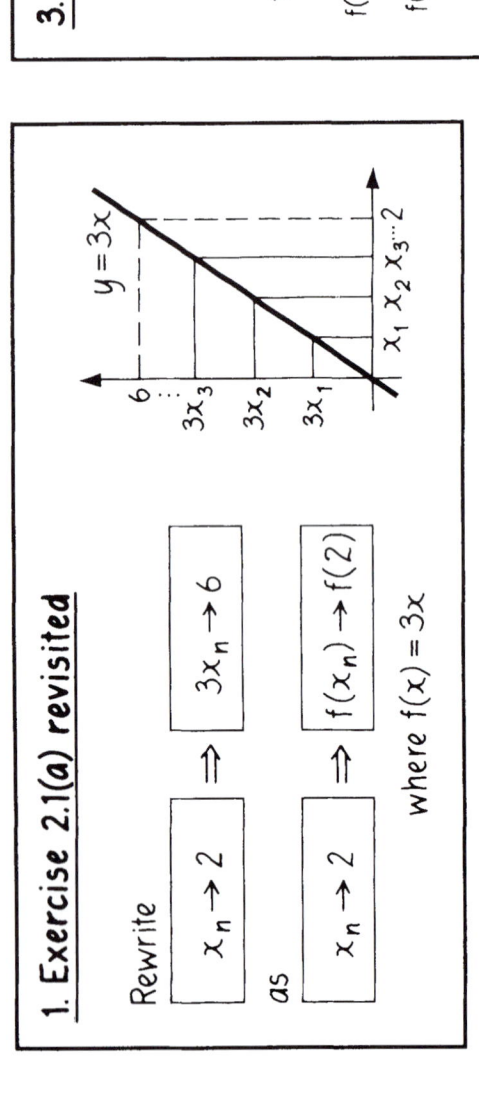

2. Exercise 2.1(b) and (c) revisited

$\boxed{x_n \to 2} \Rightarrow \boxed{x_n^2 \to 4}$

becomes

$\boxed{x_n \to 2} \Rightarrow \boxed{f(x_n) \to f(2)}$

where $f(x) = x^2$

$\boxed{\dfrac{1}{x_n} \to \dfrac{1}{2}}$

becomes

$\boxed{x_n \to 2} \Rightarrow \boxed{f(x_n) \to f(2)}$

where $f(x) = \dfrac{1}{x}$

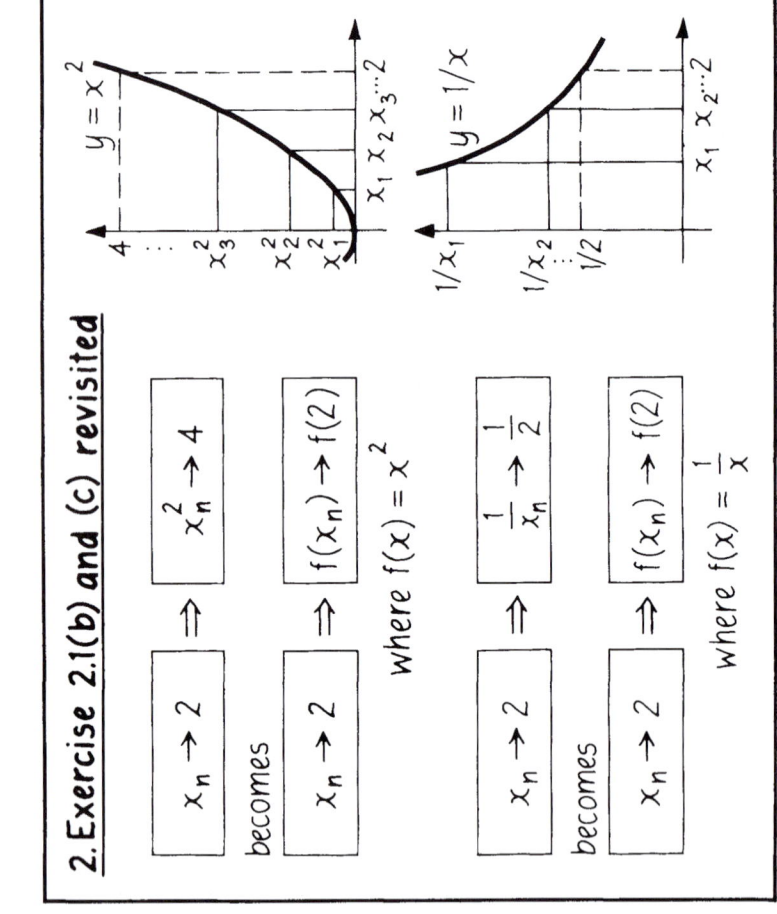

7. Strategy 2.1 Using the definition

Is $f: A \to \mathbb{R}$ continuous at a? $(a \in A)$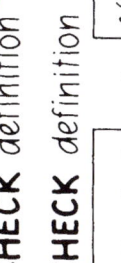

CHECK definition

GUESS behaviour........

CHECK definition HOLDS

GUESS f continuous at a

$$\boxed{x_n \to a} \implies \boxed{f(x_n) \to f(a)}$$

Frame 5

CHECK definition FAILS

GUESS f discontinuous at a

Find ONE $\{x_n\}$ in A such that

$$\boxed{x_n \to a} \text{ BUT } \boxed{f(x_n) \not\to f(a)}$$

Frame 6

---

8. Exercise 2.2

(a) Determine whether the following functions are continuous at the points given :

   (i) $f(x) = x^3 - 2x^2$, $a = 2$;

   (ii) $f(x) = [x]$, $a = 1$.

(b) Prove that these functions are continuous, i.e. continuous at every $a \in \mathbb{R}$:

   (i) $f(x) = 1$;

   (ii) $f(x) = x$.

---

5. Is $f(x) = x^3$ continuous at $\tfrac{1}{2}$?

Surely YES

$y = x^3$

WANT If $\quad x_n \to \tfrac{1}{2}$

      does $f(x_n) \to f(\tfrac{1}{2})$?

$$\boxed{x_n \to \tfrac{1}{2}} \implies \boxed{x_n^3 \to \tfrac{1}{8}}$$

KNOW This holds, by Product Rule for sequences.

RESULT f is continuous at $\tfrac{1}{2}$.

---

6. Is $f(x) = \begin{cases} 1, & x < 0 \\ 2, & x \geq 0 \end{cases}$ continuous at 0?

Surely NO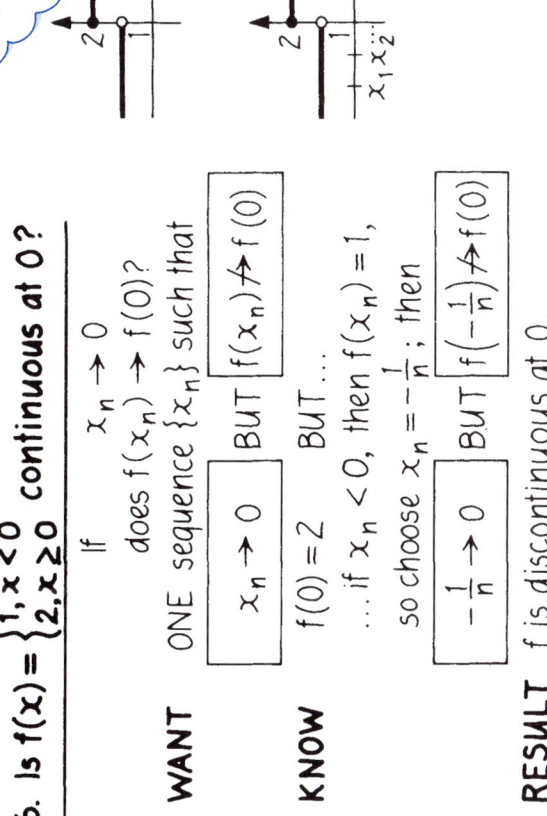

WANT If $\quad x_n \to 0$

      does $f(x_n) \to f(0)$?

ONE sequence $\{x_n\}$ such that

$$\boxed{x_n \to 0} \text{ BUT } \boxed{f(x_n) \not\to f(0)}$$

KNOW $f(0) = 2$ BUT ...

...if $x_n < 0$, then $f(x_n) = 1$,

so choose $x_n = -\tfrac{1}{n}$; then

$$\boxed{-\tfrac{1}{n} \to 0} \text{ BUT } \boxed{f(-\tfrac{1}{n}) \not\to f(0)}$$

RESULT f is discontinuous at 0.

11. Polynomials and rational functions

Use Combination Rules: build up from $f(x)=1$, $f(x)=x$

Domain of r excludes zeros of q

The following functions are continuous:

• any polynomial
$$p(x) = a_0 + a_1 x + \ldots + a_n x^n$$

• any rational function
$$r(x) = p(x)/q(x)$$

12. Is $f(x)=|x|$ continuous?

$y=|x|$

WANT
$$\boxed{x_n \to a} \;\Rightarrow\; \boxed{|x_n| \to |a|}$$

i.e.
$$\boxed{\{x_n - a\}\ \text{null}} \;\Rightarrow\; \boxed{\{|x_n| - |a|\}\ \text{null}}$$

KNOW
$$|x_n - a| \ge \big||x_n| - |a|\big| \qquad (*)$$

Triangle Inequality

so $(*)$ holds, by Squeeze Rule for null sequences

RESULT $f(x)=|x|$ is continuous.

9. Is $f(x)=\dfrac{1}{x}$ continuous?

$y = 1/x$

$\dfrac{1}{x}$ is defined everywhere EXCEPT at 0, so f has domain $\mathbb{R} - \{0\}$.

If $a \neq 0$, then
$$\boxed{x_n \to a} \;\Rightarrow\; \boxed{\dfrac{1}{x_n} \to \dfrac{1}{a}}$$

i.e.
$$\boxed{x_n \to a} \;\Rightarrow\; \boxed{f(x_n) \to f(a)}$$

so f is continuous
(at every point in its domain).

*Note
$f(x) = \begin{cases} 1/x, & x \neq 0 \\ 0, & x = 0 \end{cases}$
is discontinuous at 0*

10. Is $r(x)=(1-2x)/(x^2-4)$ continuous?

Combination Rules

If f and g are continuous at a, then so are:

the **sum** $f+g$

the **multiple** λf, $\lambda \in \mathbb{R}$

the **product** fg

the **quotient** f/g (provided $g(a) \neq 0$).

Hence $r(x) = (1-2x)/(x^2-4)$ is continuous.

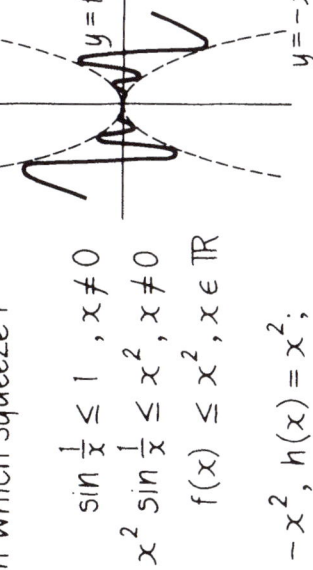

15. Squeeze Rule

Is f continuous at a?

If f, g, h are defined on an open interval I, a ∈ I, and

1. $g(x) \leq f(x) \leq h(x)$, $x \in I$
2. $g(a) = f(a) = h(a)$
3. g, h are continuous at a

then f is continuous at a.

16. Is $f(x) = \begin{cases} x^2 \sin\frac{1}{x}, & x \neq 0 \\ 0, & x = 0 \end{cases}$ continuous at 0?

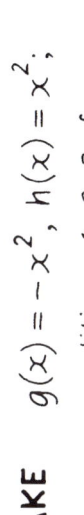

WANT g and h which squeeze f

KNOW $-1 \leq \sin\frac{1}{x} \leq 1$, $x \neq 0$

$\Rightarrow -x^2 \leq x^2 \sin\frac{1}{x} \leq x^2$, $x \neq 0$

$\Rightarrow -x^2 \leq f(x) \leq x^2$, $x \in \mathbb{R}$

TAKE $g(x) = -x^2$, $h(x) = x^2$;

conditions 1, 2, 3 of Squeeze Rule hold, with $a = 0$

RESULT f is continuous at 0.

13. Is $f(x) = \sqrt{x}$ continuous?

WANT $\boxed{x_n \to a} \ \Rightarrow \ \boxed{\sqrt{x_n} \to \sqrt{a}}$

i.e. $\boxed{\{x_n - a\} \text{ null}} \Rightarrow \boxed{\{\sqrt{x_n} - \sqrt{a}\} \text{ null}}$

KNOW $\sqrt{|x_n - a|} \geq |\sqrt{x_n} - \sqrt{a}|$ (Unit AA1, Frame 3)

$\{x_n - a\}$ null $\Rightarrow \{\sqrt{|x_n - a|}\}$ null (Power Rule)

so $\{\sqrt{x_n} - \sqrt{a}\}$ is null   (Squeeze Rule for null sequences)

RESULT $f(x) = \sqrt{x}$ is continuous.

14. Is $h(x) = \sqrt{x^2 + 1}$ continuous?

h is a composite function

$g(f(x)) = \sqrt{x^2 + 1}$ i.e. $g \circ f = h$

Composition Rule

If f is continuous at a, and g is continuous at f(a), then $g \circ f$ is continuous at a.

TAKE $f(x) = x^2 + 1$, $g(x) = \sqrt{x}$

f and g are both continuous

RESULT $h(x) = \sqrt{x^2 + 1}$ is continuous.

17

17. Is f(x) = sin x continuous at 0?

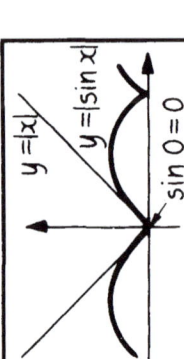

$y = |x|$
$y = |\sin x|$
$\sin 0 = 0$

WANT

KNOW $x_n \to 0 \;\Rightarrow\; \sin x_n \to 0$

$|x_n| \geq |\sin x_n|$ (∗)

so (∗) holds, by Squeeze Rule for null sequences

RESULT f(x) = sin x is continuous at 0.

[trigonometric functions are continuous]

See Subsection 2.3

19. Is f(x) = eˣ continuous at 0?

$y = \dfrac{1}{1-x}$
$y = e^x$
$y = 1+x$
$e^0 = 1$

WANT g and h which squeeze f

KNOW $1+x \leq e^x \leq \dfrac{1}{1-x}$ for $|x| < 1$

TAKE $g(x) = 1+x$, $h(x) = \dfrac{1}{1-x}$;
conditions 1, 2, 3 of
Squeeze Rule hold: I is $(-1, 1)$ and $a = 0$

RESULT $f(x) = e^x$ is continuous at 0.

[exponential functions are continuous]

18. Exercise 2.3

Prove that:

(a) $f(x) = \cos(x^2)$
is continuous;

(b) $f(x) = \begin{cases} x \sin \frac{1}{x}, & x \neq 0 \\ 0, & x = 0 \end{cases}$
is continuous at 0.

Assume that trigonometric functions are continuous.

20. Basic continuous functions

- polynomials and rational functions   Frame 11

- $f(x) = |x|$ Frame 12

- $f(x) = \sqrt{x}$ Frame 13

- trigonometric functions   Frame 17

- exponential functions   Frame 19

In the audio frames we introduced three techniques for proving that a given function f is continuous at a point a:

1. use the definition (Frame 4);
2. use the Combination Rules and the Composition Rule, together with the list of basic continuous functions (Frames 10, 14 and 20);
3. use the Squeeze Rule (Frame 15).

Normally, you should use the definition only if you think that f is *discontinuous* at a (see Strategy 2.1 in Frame 7) or if you can find no other way to prove that f is continuous at a.

To test your understanding of these techniques, try the following exercises.

Exercise 2.4   Use the Combination Rules and the Composition Rule, together with the list of basic continuous functions, to prove that the following functions are continuous (on $\mathbb{R}$).

(a) $f(x) = 7e^{-x^2}$ (b) $f(x) = x^2 + 1 + 3\sin\left(\sqrt{x^2 + 1}\,\right)$

Exercise 2.5   Determine whether the following functions are continuous at 0.

(a) $f(x) = \begin{cases} x^2 \cos(1/x^2), & x \neq 0, \\ 0, & x = 0. \end{cases}$

(b) $f(x) = \begin{cases} \sin(1/x), & x \neq 0, \\ 0, & x = 0. \end{cases}$

Hint: In part (b), use the fact that $\sin(2n + \tfrac{1}{2})\pi = 1$, for $n \in \mathbb{Z}$.

We now describe another rule for proving that a function is continuous at a point. Consider the hybrid function

$$f(x) = \begin{cases} 1 - x, & x < 0, \\ 0, & 0 \leq x \leq 1, \\ 3x - 3, & x > 1. \end{cases} \qquad (2.1)$$

> We considered this function in Unit I1, Example 3.4.

The domain of f is the whole of $\mathbb{R}$ and the graph of f looks like this.

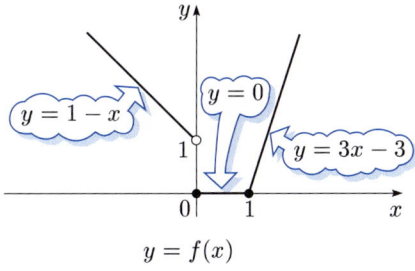

$$y = f(x)$$

From the graph, it appears that

1. f is discontinuous at $a = 0$;
2. f is continuous at all other values of a.

We can prove that f is discontinuous at 0 by using Strategy 2.1 (in Frame 7). We need to find *one* sequence $\{x_n\}$ such that

$$x_n \to 0 \quad \text{but} \quad f(x_n) \not\to f(0).$$

> The symbol $\not\to$ is read as 'does not tend to'.

We can choose

$$x_n = -\frac{1}{n}, \quad n = 1, 2, \ldots.$$

> Since $f(0)$ is defined using the rule for $[0, 1]$, we choose a simple sequence $\{x_n\}$ which tends to 0 from the left.

19

The rule for $f(x)$ which applies for $x < 0$ is $1 - x$, so

$$f(x_n) = f\left(-\frac{1}{n}\right) = 1 - \left(-\frac{1}{n}\right) = 1 + \frac{1}{n}, \quad \text{for } n = 1, 2, \ldots,$$

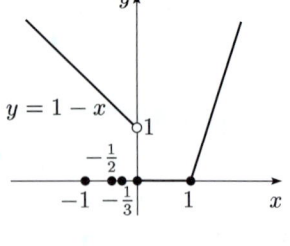

and we also have $f(0) = 0$. Hence

$$x_n \to 0 \quad \text{but} \quad f(x_n) \to 1 \neq f(0).$$

Thus f is discontinuous at 0.

But how can we prove that f *is continuous* at $a = 1$, as the graph suggests? Near 1, the graph of f consists of part of the graph $y = g(x)$ to the left of 1, glued to part of the graph $y = h(x)$ to the right of 1, where

$$g(x) = 0 \quad (x \in \mathbb{R}) \quad \text{and} \quad h(x) = 3x - 3 \quad (x \in \mathbb{R}).$$

This idea is the basis of the Glue Rule.

Glue Rule   Let f be defined on an open interval I and let $a \in I$. If there are functions g and h such that

1. $f(x) = g(x)$, for $x \in I$, $x < a$,
    $f(x) = h(x)$, for $x \in I$, $x > a$,
2. $f(a) = g(a) = h(a)$,
3. g and h are continuous at a,

then f is also continuous at a.

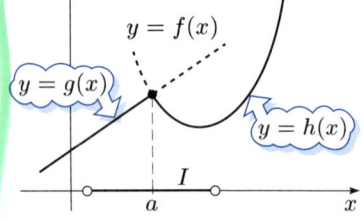

Remark   The Glue Rule does not require the functions g and h to be defined on the whole of I, though this is often the case.

We now apply the Glue Rule to the function in equation (2.1).

Example 2.1   Use the Glue Rule to prove that the function

$$f(x) = \begin{cases} 1 - x, & x < 0, \\ 0, & 0 \leq x \leq 1, \\ 3x - 3, & x > 1, \end{cases}$$

is continuous at 1.

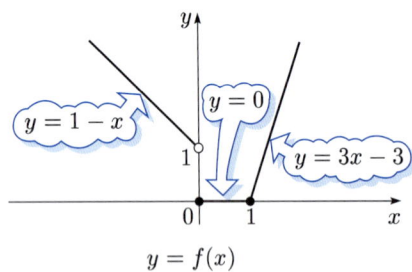
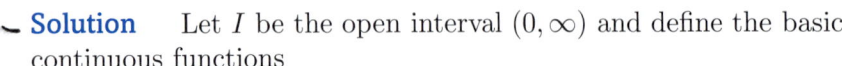

$$y = f(x)$$

Solution   Let I be the open interval $(0, \infty)$ and define the basic continuous functions

$$g(x) = 0 \quad (x \in \mathbb{R}) \quad \text{and} \quad h(x) = 3x - 3 \quad (x \in \mathbb{R}).$$

Then f is defined on I and $1 \in I$. Also,

$$f(x) = g(x), \text{ for } x \in (0, 1),$$
$$f(x) = h(x), \text{ for } x \in (1, \infty),$$

so condition 1 holds with $a = 1$.

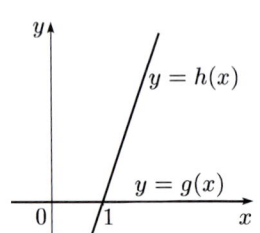

Moreover, $f(1) = g(1) = h(1) = 0$, so condition 2 holds, and g, h are continuous at 1, so condition 3 holds.

Hence f is continuous at 1, by the Glue Rule. ■

Remark   In this solution we chose $I = (0, \infty)$ because on this interval the rule of f is given by the rule of g to the left of 1 and by the rule of h to the right of 1. We could have chosen any smaller open interval that contains the point 1.

Exercise 2.6   Prove that the following function is continuous at 0.

$$f(x) = \begin{cases} e^x, & x < 0, \\ \dfrac{1}{x+1}, & x \ge 0. \end{cases}$$

There are two further situations in which we can obtain 'new continuous functions from old' and we illustrate these by examples. In both these situations, we usually take the continuity of the 'new function' for granted.

1. Consider the function f in Example 2.1. It seems evident that this function f is continuous at all points $a \in \mathbb{R} - \{0, 1\}$. For example, the continuity of f at -1 depends only on the values taken by the function f *near* the point -1, and these values are the same as those of the basic continuous function

    $$g(x) = 1 - x \quad (x \in \mathbb{R}).$$

    Since g is continuous at -1, we deduce that f is also continuous at -1. We say that continuity at a point is a *local property*. It follows similarly that f is continuous at all points $a \in \mathbb{R} - \{0, 1\}$.

2. Consider the function

    $$f(x) = \sin x \quad (x \in [-\tfrac{1}{2}\pi, \tfrac{1}{2}\pi]). \tag{2.2}$$

    The domain of this function is $[-\tfrac{1}{2}\pi, \tfrac{1}{2}\pi]$, and it certainly appears that f is continuous at each point of $[-\tfrac{1}{2}\pi, \tfrac{1}{2}\pi]$. After all, the basic continuous function

    $$g(x) = \sin x \quad (x \in \mathbb{R}),$$

    has the same rule as f, and f is the *restriction* of g to the interval $[-\tfrac{1}{2}\pi, \tfrac{1}{2}\pi]$. It follows from the definition of continuity that if a function f is the restriction of another function g, and g is continuous, then f is also continuous.

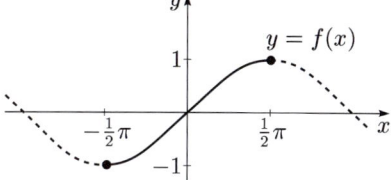

The restriction of a function was introduced in Unit I2, Section 2.

2.2 Proofs

In this subsection we give the proofs of the Combination, the Composition, the Squeeze and the Glue Rules.

First recall the Combination Rules.

If you are short of time, omit these proofs.

See Frame 10.

Combination Rules   If f and g are continuous at a, then so are

Sum Rule $f + g$;

Multiple Rule λf, for $\lambda \in \mathbb{R}$;

Product Rule fg;

Quotient Rule f/g, provided that $g(a) \ne 0$.

Proof   The proofs of these rules are similar and depend on the corresponding results for sequences. We prove only the Sum Rule.

Suppose that f and g are continuous at a. We want to deduce that $f + g$ is continuous at a.

Let the domain of f be A and the domain of g be B. Then the domain of $f + g$ is $A \cap B$ and this set contains a.

Thus, we have to show that

for each sequence $\{x_n\}$ in $A \cap B$ such that $x_n \to a$,

$$f(x_n) + g(x_n) \to f(a) + g(a). \qquad (*)$$

For $x \in A$, we have
$$(f + g)(x) = f(x) + g(x).$$

We know that $\{x_n\}$ lies in A and in B, and that both functions f and g are continuous at a. Hence

$$f(x_n) \to f(a) \quad \text{and} \quad g(x_n) \to g(a),$$

so statement $(*)$ follows by the Sum Rule for sequences. ■

Next recall the Composition Rule.

See Frame 14.

> **Composition Rule**   If f is continuous at a and g is continuous at $f(a)$, then $g \circ f$ is continuous at a.

Proof   Suppose that f is continuous at a and g is continuous at $f(a)$. We want to deduce that $g \circ f$ is continuous at a.

If f has domain A and g has domain B, then the domain of $g \circ f$ is

$$C = \{x \in A \colon f(x) \in B\}$$

and this set contains a.

Thus, we have to show that

for each sequence $\{x_n\}$ in C such that $x_n \to a$,

$$g(f(x_n)) \to g(f(a)). \qquad (*)$$

We know that $\{x_n\}$ lies in A and that f is continuous at a. Hence

$$f(x_n) \to f(a).$$

We also know that $\{f(x_n)\}$ lies in B, since $\{x_n\}$ lies in C, and that g is continuous at $f(a)$. Hence $g(f(x_n)) \to g(f(a))$, so statement $(*)$ is true. ■

Next recall the Squeeze Rule.

See Frame 15.

> **Squeeze Rule**   Let f, g and h be defined on an open interval I and let $a \in I$. If
>
> 1. $g(x) \le f(x) \le h(x)$, for $x \in I$,
> 2. $g(a) = f(a) = h(a)$,
> 3. g and h are continuous at a,
>
> then f is also continuous at a.

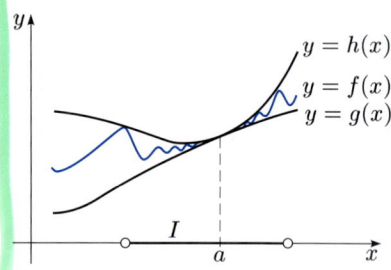

Proof   Suppose that f, g and h satisfy the conditions of the theorem. We want to prove that f is continuous at a.

Thus, we have to show that

for each sequence $\{x_n\}$ in the domain of f such that $x_n \to a$,

$$f(x_n) \to f(a). \tag{$*$}$$

Since $x_n \to a$ and I is an open interval, there is an integer N such that

$$x_n \in I, \quad \text{for } n > N.$$

Hence, by condition 1,

$$g(x_n) \le f(x_n) \le h(x_n), \quad \text{for } n > N.$$

By conditions 2 and 3,

$$\lim_{n\to\infty} g(x_n) = \lim_{n\to\infty} h(x_n) = f(a),$$

so statement $(*)$ follows, by the Squeeze Rule for sequences. ■

Here we are using the definition of a convergent sequence; see Unit AA2, Section 3.

See Unit AA2, Section 3.

Finally, we prove the Glue Rule.

See page 20.

> **Glue Rule**   Let f be defined on an open interval I and let $a \in I$.
> If there are functions g and h such that
>
> 1. $f(x) = g(x)$, for $x \in I$, $x < a$,
> $f(x) = h(x)$, for $x \in I$, $x > a$,
> 2. $g(a) = f(a) = h(a)$,
> 3. g and h are continuous at a,
>
> then f is also continuous at a.

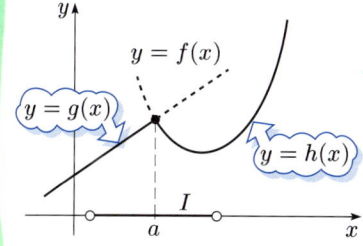

Proof   Suppose that f, g and h satisfy the conditions of the theorem. We want to prove that

for each sequence $\{x_n\}$ in the domain of f such that $x_n \to a$,

$$f(x_n) \to f(a). \tag{$*$}$$

Since $x_n \to a$ and I is an open interval, there is an integer N such that

$$x_n \in I, \quad \text{for } n \ge N.$$

Then $\{x_n\}_N^\infty$ consists of two subsequences $\{x_{m_k}\}$ and $\{x_{n_k}\}$ defined by the conditions

$$x_{m_k} < a, \quad \text{for } k = 1, 2, \ldots, \quad \text{and} \quad x_{n_k} \ge a, \quad \text{for } k = 1, 2, \ldots.$$

By conditions 1 and 3, we have

$$g(x_{m_k}) \to g(a) \text{ as } k \to \infty \quad \text{and} \quad h(x_{n_k}) \to h(a) \text{ as } k \to \infty.$$

Hence, by conditions 1 and 2, we have

$$f(x_{m_k}) \to f(a) \text{ as } k \to \infty \quad \text{and} \quad f(x_{n_k}) \to f(a) \text{ as } k \to \infty.$$

Since both subsequences of $\{f(x_n)\}$ tend to $f(a)$, statement $(*)$ follows. ■

Recall that $\{x_n\}_N^\infty$ is the sequence

$$x_N, x_{N+1}, \ldots.$$

See Unit AA2, Theorem 4.3.

2.3 Trigonometric and exponential functions

In Frames 17 and 19 of the audio we used inequalities for the sine and exponential functions to prove that these functions are continuous at 0. We now prove these inequalities, and deduce that the trigonometric functions and the exponential function are continuous at all points of $\mathbb{R}$.

First we obtain a fundamental inequality for the sine function.

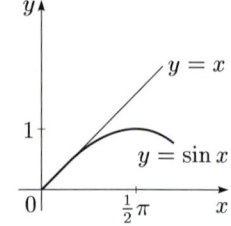

Sine Inequality

$\sin x \leq x, \quad$ for $0 \leq x \leq \frac{1}{2}\pi$.

Proof   If $x = 0$, then $\sin 0 = 0$, so there is equality.

Suppose next that $0 < x \leq \frac{1}{2}\pi$, and consider the following diagram, which represents a quarter circle, centred at the origin, with radius 1.

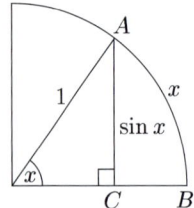

Since the circle has radius 1, the arc AB has length x and the perpendicular AC has length $\sin x$. Hence

$\sin x \leq x, \quad$ for $0 < x \leq \frac{1}{2}\pi.$ ■

We can now deduce the inequality used in Frame 17 of the audio.

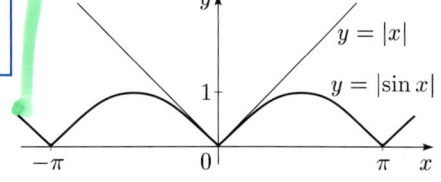

Corollary $|\sin x| \leq |x|$, for $x \in \mathbb{R}$.

Proof   The Sine Inequality shows that this inequality holds for $0 \leq x \leq \frac{1}{2}\pi$. For $x > \frac{1}{2}\pi$, we have

$|\sin x| \leq 1 < \frac{1}{2}\pi < x = |x|,$

so the inequality is also true in this case.

Finally, the inequality holds for $x < 0$, since

$|\sin(-x)| = |\sin x| \quad$ and $\quad |-x| = |x|.$ ■

Next we obtain two fundamental inequalities for the exponential function.

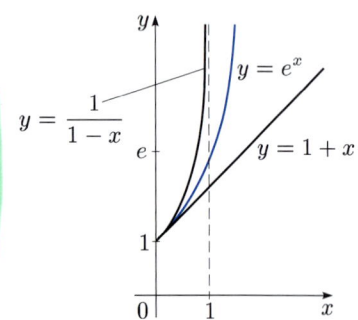

Exponential Inequalities

(a) $e^x \geq 1 + x$, for $x \geq 0$;

(b) $e^x \leq \dfrac{1}{1-x}$, for $0 \leq x < 1$.

Proof   We prove both inequalities using the exponential series

$$e^x = 1 + x + \frac{x^2}{2!} + \frac{x^3}{3!} + \cdots, \quad \text{for } x \geq 0.$$

See Unit AA3, Subsection 4.1.

(a) For $x \geq 0$, we have $x^2/2! \geq 0$, $x^3/3! \geq 0$, and so on. Hence

$$e^x \geq 1 + x, \quad \text{for } x \geq 0.$$

(b) For $x \geq 0$, we also have $x^2/2! \leq x^2$, $x^3/3! \leq x^3$, and so on. Hence

$$e^x \leq 1 + x + x^2 + x^3 + \cdots.$$

The series on the right is a geometric series, which is convergent with sum $1/(1-x)$, for $0 \leq x < 1$. Hence

See Unit AA3, Subsection 1.1.

$$e^x \leq \frac{1}{1-x}, \quad \text{for } 0 \leq x < 1. \quad \blacksquare$$

We can now deduce the inequalities used in Frame 19 of the audio.

> **Corollary** $1 + x \leq e^x \leq \dfrac{1}{1-x}$, for $|x| < 1$.

Proof   The Exponential Inequalities show that these inequalities both hold for $0 \leq x < 1$. For $-1 < x < 0$, we have $0 < -x < 1$, so

$$1 + (-x) \leq e^{-x} \leq \frac{1}{1-(-x)}.$$

Taking reciprocals and reversing these inequalities (which is possible since all three expressions are positive for $-1 < x < 0$), we obtain

$$1 + x \leq e^x \leq \frac{1}{1-x}, \quad \text{for } -1 < x < 0.$$

Hence

$$1 + x \leq e^x \leq \frac{1}{1-x}, \quad \text{for } |x| < 1. \quad \blacksquare$$

Now we can show that the sine function and the exponential function are continuous everywhere.

> **Theorem 2.1**   The following functions are continuous:
>
> (a) the trigonometric functions (sine, cosine and tangent);
>
> (b) the exponential function.

Proof

(a) To prove that the sine function is continuous at $a \in \mathbb{R}$, we want to show that

for each sequence $\{x_n\}$ in $\mathbb{R}$ such that $x_n \to a$,

$$\sin x_n \to \sin a. \qquad (*)$$

To do this, we use the trigonometric identity

$$\sin x - \sin a = 2 \cos \left(\tfrac{1}{2}(x+a)\right) \sin \left(\tfrac{1}{2}(x-a)\right).$$

We obtain

$$
\begin{aligned}
&|\sin x_n - \sin a| \\
&= \left|2 \cos \left(\tfrac{1}{2}(x_n+a)\right) \sin \left(\tfrac{1}{2}(x_n-a)\right)\right| \\
&\le 2\left|\sin \left(\tfrac{1}{2}(x_n-a)\right)\right| \quad (\text{since } |\cos x| \le 1) \\
&\le 2\left|\tfrac{1}{2}(x_n-a)\right| \quad (\text{by the corollary to the Sine Inequality}) \\
&= |x_n - a|.
\end{aligned}
$$

Thus, if $\{x_n - a\}$ is null, then $\{\sin x_n - \sin a\}$ is null, by the Squeeze Rule for sequences, so statement $(*)$ holds.

The continuity of the cosine and tangent functions now follows from the identities

$$\cos x = \sin \left(x + \tfrac{1}{2}\pi\right) \quad \text{and} \quad \tan x = \frac{\sin x}{\cos x},$$

using the Composition Rule and the Quotient Rule.

(b) To prove that the exponential function is continuous at $a \in \mathbb{R}$, we want to show that

for each sequence $\{x_n\}$ in $\mathbb{R}$ such that $x_n \to a$,

$$e^{x_n} \to e^a. \tag{$*$}$$

Now, if $\{x_n - a\}$ is null, then there is a positive integer N such that $|x_n - a| < 1$, for $n > N$. Applying the corollary to the Exponential Inequalities, with $x_n - a$ instead of x, we obtain

$$1 + (x_n - a) \le e^{x_n - a} \le \frac{1}{1 - (x_n - a)}, \quad \text{for } n > N.$$

Thus $e^{x_n - a} \to 1$, by the Squeeze Rule for sequences. Hence $e^{x_n} = e^a e^{x_n - a} \to e^a$, so statement $(*)$ holds. ∎

This identity can be obtained by writing
$$x = \tfrac{1}{2}(x+a) + \tfrac{1}{2}(x-a)$$
and
$$a = \tfrac{1}{2}(x+a) - \tfrac{1}{2}(x-a),$$
and then expanding $\sin x$ and $\sin a$.

Further exercises

Exercise 2.7   Use the appropriate rules, together with the list of basic continuous functions, to prove that the following functions are continuous.

(a) $f(x) = \exp(\sin(x^2 + 1)) \quad (x \in \mathbb{R})$

(b) $f(x) = e^{\sqrt{x}} + x^5 \quad (x \in [0, \infty))$

Exercise 2.8   Determine whether the following functions are continuous at 0.

(a) $f(x) = \begin{cases} x, & x < 0, \\ \dfrac{1}{x+1}, & x \ge 0. \end{cases}$

(b) $f(x) = \begin{cases} \sin x \sin(1/x), & x \ne 0, \\ 0, & x = 0. \end{cases}$

(c) $f(x) = \begin{cases} (1/x) \cos(1/x), & x \ne 0, \\ 0, & x = 0. \end{cases}$

(d) $f(x) = \begin{cases} 0, & x < 0, \\ \sqrt{x}, & x \ge 0. \end{cases}$

Exercise 2.9   Prove that the following function is continuous at $\frac{1}{2}\pi$ and at $-\frac{1}{2}\pi$.

$$f(x) = \begin{cases} -1, & x \leq -\frac{1}{2}\pi, \\ \sin x, & -\frac{1}{2}\pi < x < \frac{1}{2}\pi, \\ 1, & x \geq \frac{1}{2}\pi. \end{cases}$$

3   Properties of continuous functions

After working through this section, you should be able to:

(a) state the Intermediate Value Theorem and use it to prove that certain equations have solutions;

(b) determine an interval which contains all the *zeros* of a given polynomial;

(c) state the Extreme Value Theorem and the Boundedness Theorem.

In this section, which includes the video section, we describe some of the fundamental properties of continuous functions, and we see that these properties hold for continuous functions defined on *bounded closed intervals*; that is, intervals of the form $[a, b]$. We say that a function f is **continuous on** an interval I if f is continuous at each point of I.

3.1 Intermediate Value Theorem

At the end of Section 1, we considered the function

$$f(x) = x^5 + x - 1 \quad (x \in \mathbb{R})$$

and we pointed out that it is not easy to prove that $f(\mathbb{R}) = \mathbb{R}$.

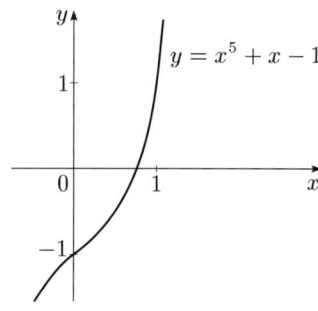

For example, is there a value of x such that $f(x) = 0$? In other words, is there a solution of the equation

$$x^5 + x - 1 = 0?$$

The shape of the graph $y = x^5 + x - 1$ certainly suggests that such a number x exists. Since $f(0) = -1$ and $f(1) = 1$, we expect there to be some number x in the interval $(0, 1)$ such that $f(x) = 0$. However, we do not have a formula for solving the above equation to find x.

The key to proving that such a number x exists lies in the fact that f is a continuous function, so there cannot be any gaps in its graph. We can prove this by using the *Intermediate Value Theorem*.

Theorem 3.1 Intermediate Value Theorem

Let f be a function continuous on $[a, b]$ and let k be any number lying between $f(a)$ and $f(b)$. Then there exists a number c in (a, b) such that

$$f(c) = k.$$

We have either

$f(a) < k < f(b)$, or

$f(a) > k > f(b)$.

27

This result is illustrated below in the two possible cases.

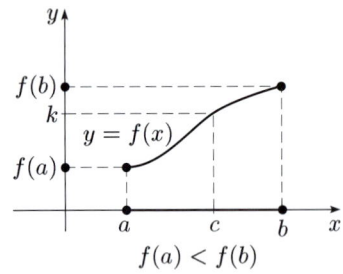

$$f(a) < f(b)$$

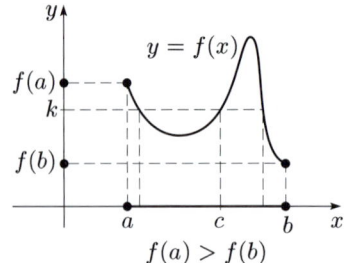

$$f(a) > f(b)$$

As the graph on the right shows, there may be more than one possible value of c such that $f(c) = k$.

Remark   The conclusion of the Intermediate Value Theorem may be *false* if f is discontinuous at even *one* point of $[a, b]$. For example, the function

$$f(x) = \begin{cases} 1/x, & 0 < |x| \le 1, \\ 0, & x = 0, \end{cases} \qquad (3.1)$$

is continuous on $[-1, 1]$ except at 0. For this function,

$$f(-1) = -1 \quad \text{and} \quad f(1) = 1,$$

but there is no number c in $(-1, 1)$ such that $f(c) = \frac{1}{2}$.

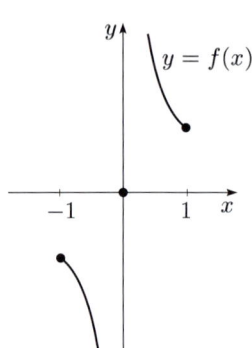

Here is a typical application of the Intermediate Value Theorem.

Example 3.1   Use the Intermediate Value Theorem to prove that there is a number c in $(0, 1)$ such that

$$c^5 + c - 1 = 0.$$

Solution   Consider the basic continuous function

$$f(x) = x^5 + x - 1.$$

Then f is continuous on $[0, 1]$ and also

$$f(0) = -1 \quad \text{and} \quad f(1) = 1.$$

Since

$$f(0) < 0 < f(1),$$

it follows from the Intermediate Value Theorem that there is a number c in $(0, 1)$ such that

$$f(c) = 0; \quad \text{that is,} \quad c^5 + c - 1 = 0. \quad \blacksquare$$

The function f is strictly increasing on $[0, 1]$, so in this case the number c must be *unique*.

To obtain further information about the location of the number c in Example 3.1, we can use, for example, the *bisection method*. This involves repeatedly bisecting the interval containing the solution and testing the values of f at the bisection points, in order to find shorter and shorter intervals in which the number c must lie.

For example, at the first bisection the function

$$f(x) = x^5 + x - 1$$

satisfies

$$f\left(\tfrac{1}{2}\right) = \left(\tfrac{1}{2}\right)^5 + \tfrac{1}{2} - 1 < 0 \quad \text{and} \quad f(1) = 1 > 0,$$

so the number c must lie in $(\tfrac{1}{2}, 1)$. To find an interval of length $\tfrac{1}{4}$ containing c, we next consider the value $f(\tfrac{3}{4})$, and so on.

— **Exercise 3.1**   Use the bisection method to find an interval of length $\frac{1}{16}$ containing the number c such that

$$c^5 + c - 1 = 0.$$

The bisection method can also be used to prove the Intermediate Value Theorem. This is one of the topics covered in the video programme.

Watch the video programme 'The Intermediate Value Theorem'.

− 3.2 Review of the video programme

The programme begins by considering the temperatures at various points on the Earth's equator. We ask whether there must always be a pair of *antipodal points* (that is, points which lie at opposite ends of a line segment through the centre of the Earth) on the equator at which the temperature is the same. The following result uses the Intermediate Value Theorem to show that the answer is 'yes'. Here $g(\theta)$ represents the temperature at a point on the equator at an angle θ radians east of the Greenwich meridian.

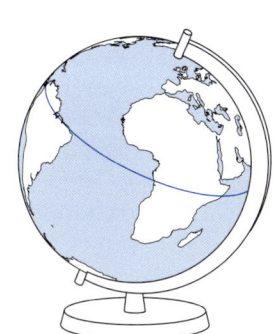

Theorem 3.2 Antipodal Points Theorem

If $g: [0, 2\pi] \longrightarrow \mathbb{R}$ is a continuous function and $g(0) = g(2\pi)$, then there exists a number c in $[0, \pi]$ such that

$$g(c) = g(c + \pi).$$

If
$$g(c) = g(c + \pi),$$
then c and $c + \pi$ represent antipodal points with the same temperature.

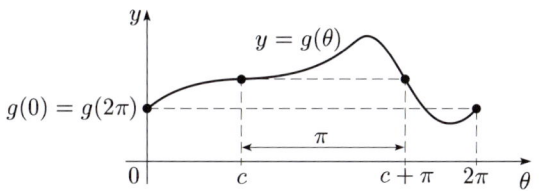

Proof   First note that if $g(0) = g(\pi)$, then we can take $c = 0$. So let us assume that

$$g(0) < g(\pi). \tag{3.2}$$

Now we define

$$h(\theta) = g(\theta + \pi) \quad (\theta \in [0, \pi]),$$

and consider the graphs $y = g(\theta)$ and $y = h(\theta)$, for $0 \le \theta \le \pi$.

The proof in the case $g(0) > g(\pi)$ is similar.

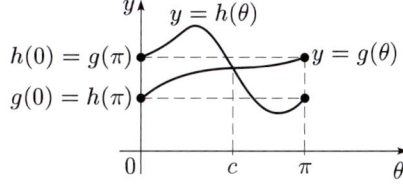

The graph $y = h(\theta)$ is obtained by translating to the left the part of the graph $y = g(\theta)$ corresponding to $\pi \le \theta \le 2\pi$.

Since

$$h(0) = g(0 + \pi) = g(\pi) \quad \text{and} \quad h(\pi) = g(\pi + \pi) = g(0),$$

inequality (3.2) can be rewritten as

$$g(0) < h(0) \quad \text{and} \quad g(\pi) > h(\pi).$$

This suggests that our two graphs must cross at some point c in $(0, \pi)$, giving

$$g(c) = h(c) \quad \text{and hence} \quad g(c) = g(c + \pi).$$

To make this argument rigorous, we define a function f as

$$f(\theta) = g(\theta) - h(\theta) \quad (\theta \in [0, \pi]),$$

which is continuous on $[0, \pi]$. Also,

$$f(0) = g(0) - h(0) < 0 \quad \text{and} \quad f(\pi) = g(\pi) - h(\pi) > 0.$$

Thus, by the Intermediate Value Theorem with $k = 0$, there exists a number c in $(0, \pi)$ such that $f(c) = 0$, so $g(c) = h(c)$ and hence

$$g(c) = g(c + \pi),$$

as required. ∎

Next in the programme, we prove the following special case of the Intermediate Value Theorem.

The general case can be deduced from this special case ($k = 0$) by considering the function
$$F(x) = f(x) - k.$$

Intermediate Value Theorem (special case)

Let f be a function continuous on $[a, b]$ and suppose that

$$f(a) < 0 < f(b).$$

Then there exists a number c in (a, b) such that

$$f(c) = 0.$$

Proof   We use the bisection method.

If you are short of time, omit this proof.

First we define $[a_0, b_0] = [a, b]$ and $p = \frac{1}{2}(a_0 + b_0)$. If $f(p) = 0$, then the proof is complete, since we can take $c = p$. Otherwise, we define

$$[a_1, b_1] = \begin{cases} [a_0, p], & \text{if } f(p) > 0, \\ [p, b_0], & \text{if } f(p) < 0. \end{cases}$$

In either case, we have
1. $[a_1, b_1] \subseteq [a_0, b_0]$;
2. $b_1 - a_1 = \frac{1}{2}(b_0 - a_0)$;
3. $f(a_1) < 0 < f(b_1)$.

Now repeat this process, bisecting $[a_1, b_1]$ to obtain $[a_2, b_2]$, and so on. If, at any stage, we encounter a bisection point p such that $f(p) = 0$, then the proof is complete. Otherwise, we obtain a sequence of closed intervals

$$[a_n, b_n], \quad n = 0, 1, 2, \ldots,$$

with the properties
1. $[a_{n+1}, b_{n+1}] \subseteq [a_n, b_n]$, for $n = 0, 1, 2, \ldots$;
2. $b_n - a_n = \left(\frac{1}{2}\right)^n (b_0 - a_0)$, for $n = 0, 1, 2, \ldots$;
3. $f(a_n) < 0 < f(b_n)$, for $n = 0, 1, 2, \ldots$.

Property 1 implies that $\{a_n\}$ is increasing and bounded above by b_0. Hence, by the Monotone Convergence Theorem, $\{a_n\}$ is convergent. Let

$$\lim_{n \to \infty} a_n = c.$$

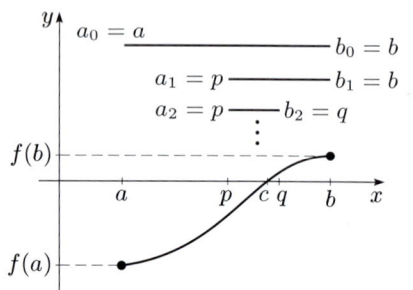

See Unit AA2, Theorem 5.1.

Then, by property 2 and the Combination Rules for sequences,

See Unit AA2, Section 3.

$$\lim_{n \to \infty} b_n = \lim_{n \to \infty} \left(a_n + (b_n - a_n) \right)$$

$$= \lim_{n \to \infty} a_n + \lim_{n \to \infty} \left(\tfrac{1}{2} \right)^n (b_0 - a_0)$$

$$= c + 0 = c.$$

Now we use the fact that f is continuous at c to obtain

$$\lim_{n \to \infty} f(a_n) = f(c) \quad \text{and} \quad \lim_{n \to \infty} f(b_n) = f(c).$$

By property 3, $f(a_n) < 0$, for $n = 0, 1, 2, \ldots$, so $f(c) \leq 0$, by the Limit Inequality Rule. Likewise, $f(c) \geq 0$ because $f(b_n) > 0$, for $n = 0, 1, 2, \ldots$. Hence $f(c) = 0$, as required. ■

See Unit AA2, Section 3.

The programme ends with some unexpected applications of the Intermediate Value Theorem (Theorem 3.1) to square tables and to pancakes.

3.3 Post-programme work: locating zeros

If f is a function and c is a real number such that

$$f(c) = 0,$$

then c is called a **zero** of the function f. We sometimes say that the function **vanishes** at c.

In the analysis units we are interested in *real* zeros of functions; we do not consider complex zeros.

We often show that an equation has a solution by proving that a related continuous function has a zero (using the Intermediate Value Theorem with $k = 0$).

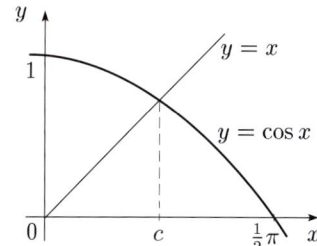

Exercise 3.2   Prove that the equation

$$\cos x = x$$

has a solution in the interval $(0, 1)$.

Hint: Consider the function $f(x) = \cos x - x$.

Exercise 3.3   Suppose that the function $f \colon [0, 1] \longrightarrow [0, 1]$ is continuous. Prove that the equation

$$f(x) = x$$

has a solution in the interval $[0, 1]$.

Hint: Consider the function $g(x) = f(x) - x \ (x \in [0, 1])$.

Zeros of polynomials

We now briefly consider the problem of locating the zeros of polynomial functions. First try the following exercise.

Zeros of polynomials were discussed in Unit I2, Section 4, and Unit I3, Section 2. In particular, recall that a polynomial of degree n has at most n zeros.

Exercise 3.4   Let

$$p(x) = x^6 - 4x^4 + x + 1 \quad (x \in \mathbb{R}).$$

Prove that p has a zero in each of the intervals $(-1, 0), (0, 1)$ and $(1, 2)$.

When we wish to locate the zeros (if any) of a given polynomial, we can begin by applying the following result. This gives an interval in which the zeros *must* lie.

Theorem 3.3 Let

$$p(x) = x^n + a_{n-1}x^{n-1} + \cdots + a_1 x + a_0 \quad (x \in \mathbb{R}),$$

where $a_0, a_1, \ldots, a_{n-1} \in \mathbb{R}$. Then all the zeros of p (if there are any) lie in the open interval $(-M, M)$, where

$$M = 1 + \max\{|a_{n-1}|, \ldots, |a_1|, |a_0|\}.$$

Note that the coefficient of x^n is 1.

We ask you to prove Theorem 3.3 in Exercise 3.9.

Example 3.2   Prove that the following polynomial has at least two zeros:

$$p(x) = x^4 - 2x^2 - x + 1 \quad (x \in \mathbb{R}).$$

Solution   Since

$$M = 1 + \max\{|-2|, |-1|, |1|\} = 3,$$

it follows from Theorem 3.3 that all the zeros of p lie in $(-3, 3)$.

We now compile a table of values of $p(n)$, for integers n in $[-3, 3]$.

n	-3	-2	-1	0	1	2	3
$p(n)$	67	11	1	1	-1	7	61

Often, it is not necessary to compute the values of $p(n)$ for *all* integers n in $[-M, M]$.

We find that $p(0)$ and $p(1)$ have opposite signs, as do $p(1)$ and $p(2)$. Thus, since p is continuous, it has a zero in each of the open intervals $(0, 1)$ and $(1, 2)$, by the Intermediate Value Theorem.

Thus we have proved that p has *at least* two zeros. ■

This function p has *exactly* two zeros.

Exercise 3.5   Prove that the following polynomial has at least three zeros:

$$p(x) = x^5 + 3x^4 - x - 1 \quad (x \in \mathbb{R}).$$

3.4 Extreme Value Theorem

We now describe another important property of continuous functions. First we give the following definitions.

Definition   Let f be a function with domain A. Then

f has **maximum value** $f(c)$ in A if $c \in A$ and

$$f(x) \le f(c), \quad \text{for } x \in A;$$

f has **minimum value** $f(c)$ in A if $c \in A$ and

$$f(c) \le f(x), \quad \text{for } x \in A;$$

f is **bounded** on A if, for some $M \in \mathbb{R}$,

$$|f(x)| \le M, \quad \text{for } x \in A.$$

An **extreme value** is a maximum or minimum value.

For example, the function $f(x) = \sin x$, with domain $\mathbb{R}$, has maximum value 1 in $\mathbb{R}$, since

$$\sin x \le 1 = \sin \tfrac{1}{2}\pi, \quad \text{for } x \in \mathbb{R}.$$

Also, this function f is bounded on $\mathbb{R}$, since

$$|\sin x| \le 1, \quad \text{for } x \in \mathbb{R}.$$

On the other hand, the function defined in equation (3.1) is not bounded on its domain $[-1, 1]$.

Note that a function can be bounded without having a maximum value or a minimum value; for example, the function

$$g(x) = \frac{x}{1 + |x|}$$

is bounded on its domain $\mathbb{R}$, since

$$|g(x)| = \frac{|x|}{1 + |x|} < 1, \quad \text{for } x \in \mathbb{R},$$

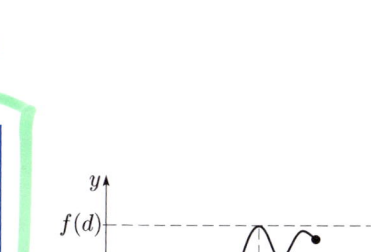

but g is strictly increasing on $\mathbb{R}$, so it has no maximum or minimum value on $\mathbb{R}$.

The following result states that a continuous function on a bounded closed interval always has a maximum value and a minimum value.

Theorem 3.4 Extreme Value Theorem

Let f be a function continuous on $[a, b]$. Then there exist numbers c and d in $[a, b]$ such that

$$f(c) \leq f(x) \leq f(d), \quad \text{for } x \in [a, b].$$

The following consequence of the Extreme Value Theorem states that, if f is continuous on $[a, b]$, then f is bounded on $[a, b]$.

Theorem 3.5 Boundedness Theorem

Let f be a function continuous on $[a, b]$. Then there exists a number M such that

$$|f(x)| \leq M, \quad \text{for } x \in [a, b].$$

Remark   The function defined in equation (3.1) shows that the conclusions of the Extreme Value Theorem and the Boundedness Theorem may be *false* if f is discontinuous at even *one* point of $[a, b]$.

Proof of the Extreme Value Theorem

We prove that there exists a number d in $[a, b]$ such that

$$f(x) \leq f(d), \quad \text{for } x \in [a, b].$$

We use the function

$$g(x) = \frac{x}{1 + |x|},$$

which is strictly increasing and continuous on $\mathbb{R}$, with

$$|g(x)| < 1, \quad \text{for } x \in \mathbb{R}.$$

Then the function

$$h(x) = g(f(x)) \quad (x \in [a, b])$$

is continuous, by the Composition Rule, and

$$|h(x)| < 1, \quad \text{for } x \in [a, b].$$

If you are short of time, omit this proof.

Similar reasoning shows that there exists $c \in [a, b]$ such that
$$f(c) \leq f(x), \quad \text{for } x \in [a, b].$$

33

Hence the image $h([a, b])$ is bounded. Thus, by the Least Upper Bound Property of $\mathbb{R}$,

See Unit AA1, Subsection 4.3.

the supremum, M say, of $h([a, b])$ exists.

We now use the bisection method to find $d \in [a, b]$ such that $h(d) = M$.

We define $[a_0, b_0] = [a, b]$ and $p = \frac{1}{2}(a_0 + b_0)$. Then at least one of the image sets $h([a_0, p])$ and $h([p, b_0])$ must have least upper bound M. Thus we can choose $[a_1, b_1]$ such that

If both $h([a_0, p])$ and $h([p, b_0])$ have upper bounds less than M, then $h([a, b])$ also has an upper bound less than M.

1. $[a_1, b_1] \subseteq [a_0, b_0]$;
2. $b_1 - a_1 = \frac{1}{2}(b_0 - a_0)$;
3. $M = \sup h([a_1, b_1])$.

Now we repeat this process, bisecting $[a_1, b_1]$ to obtain $[a_2, b_2]$, and so on. We obtain a sequence of closed intervals

$$[a_n, b_n], \quad n = 0, 1, 2, \ldots,$$

with the properties

1. $[a_{n+1}, b_{n+1}] \subseteq [a_n, b_n]$, for $n = 0, 1, 2, \ldots$;
2. $b_n - a_n = \left(\frac{1}{2}\right)^n (b_0 - a_0)$, for $n = 0, 1, 2, \ldots$;
3. $M = \sup h([a_n, b_n])$, for $n = 0, 1, 2, \ldots$.

As in the proof of the Intermediate Value Theorem, properties 1 and 2 imply that there is a real number $d \in [a, b]$ such that

$$\lim_{n \to \infty} a_n = \lim_{n \to \infty} b_n = d.$$

By property 3, for each $n = 1, 2, \ldots$, there is a number t_n such that

This holds because each number $M - 1/n$ is *not* an upper bound of the image $h([a, b])$.

$$a_n \le t_n \le b_n \quad \text{and} \quad M - 1/n \le h(t_n) \le M.$$

Hence, by the Squeeze Rule for sequences,

$$\lim_{n \to \infty} t_n = d \quad \text{and} \quad \lim_{n \to \infty} h(t_n) = M.$$

Thus, by the continuity of h at d,

$$h(d) = \lim_{n \to \infty} h(t_n) = M,$$

so

$$h(x) = g(f(x)) \le g(f(d)) = h(d), \quad \text{for } x \in [a, b].$$

Since g is strictly increasing, it follows that

$$f(x) \le f(d), \quad \text{for } x \in [a, b],$$

as required. ∎

Further exercises

Exercise 3.6   Prove that each of the following polynomials has the stated number of zeros:

(a) $p(x) = x^4 - 4x^3 + 3x^2 + 2x - 1$,   four zeros;

(b) $p(x) = 3x^3 - 8x^2 + x + 3$,   three zeros.

Exercise 3.7   Prove that the function

$$f(x) = x - \sin x - \tfrac{2}{3}\pi \quad (x \in \mathbb{R})$$

has a zero in $\left(\frac{2}{3}\pi, \frac{5}{6}\pi\right)$.

Exercise 3.8 Suppose that the function $f\colon [0,1] \longrightarrow [0,1]$ is continuous. Prove that the equation

$$f(x) = x^3$$

has a solution in the interval $[0,1]$.

Hint: Consider the function $g(x) = f(x) - x^3$ $(x \in [0,1])$.

Exercise 3.9 (Harder) Let

$$p(x) = x^n + a_{n-1}x^{n-1} + \cdots + a_1x + a_0 \quad (x \in \mathbb{R}),$$

and

$$r(x) = \frac{p(x)}{x^n} - 1 = \frac{a_{n-1}}{x} + \cdots + \frac{a_1}{x^{n-1}} + \frac{a_0}{x^n} \quad (x \in \mathbb{R} - \{0\}).$$

(a) Prove that

$$|r(x)| < \frac{K}{|x| - 1}, \quad \text{for } |x| > 1,$$

where $K = \max\{|a_{n-1}|, \ldots, |a_1|, |a_0|\}$.

Deduce that

$$|r(x)| < 1, \quad \text{for } |x| \geq M,$$

where $M = 1 + \max\{|a_{n-1}|, \ldots, |a_1|, |a_0|\}$.

(b) Use part (a) to prove that $p(x)$ has the same sign as x^n for $|x| \geq M$, and deduce that any zero of p must lie in $(-M, M)$.

(c) Use part (b) to prove that, if n is odd, then p has at least one zero.

(d) Use part (b) to prove that, if n is even, then p has a minimum value; that is, there exists a number c such that

$$p(x) \geq p(c), \quad \text{for } x \in \mathbb{R}.$$

Hint: Consider the function $q(x) = p(x) - a_0$. Show that q has a minimum value on $\mathbb{R}$, taken by q at some point in $[-M, M]$.

Exercise 3.9(a) and (b) form the proof of Theorem 3.3.

4   Inverse functions

After working through this section, you should be able to:

(a) use the Inverse Function Rule to establish that a given function $f\colon I \longrightarrow J$ has a continuous inverse function $f^{-1}\colon J \longrightarrow I$;

(b) define the inverse functions of certain standard functions;

(c) define a^x for $a > 0$ and any $x \in \mathbb{R}$.

4.1 Inverse Function Rule

At the end of Section 1 we discussed the function

$$f(x) = x^5 + x - 1 \quad (x \in \mathbb{R}).$$

We showed that f is strictly increasing and hence one-one, but we could not prove that $f(\mathbb{R}) = \mathbb{R}$, so we could not prove that the inverse function f^{-1} has domain $\mathbb{R}$.

Now, however, we know that f is a continuous function, so the Intermediate Value Theorem can be used to prove that there is a point c in $(0,1)$ such that

$$f(c) = 0.$$

Similarly, we can prove that, for each $y \in \mathbb{R}$, the equation

$$f(x) = y$$

has a solution x. This means that $f(\mathbb{R}) = \mathbb{R}$, so f^{-1} has domain $\mathbb{R}$.

Moreover, this inverse function f^{-1} is also continuous, by the Inverse Function Rule.

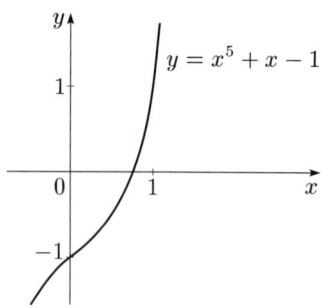

See Example 3.1.

The proof of the Inverse Function Rule is in Subsection 4.4.

Inverse Function Rule   Let $f \colon I \longrightarrow J$, where I is an interval and J is the image $f(I)$, be a function such that

1. f is strictly increasing on I;
2. f is continuous on I.

Then J is an interval and f has an inverse function $f^{-1} \colon J \longrightarrow I$ such that

1'. f^{-1} is strictly increasing on J;
2'. f^{-1} is continuous on J.

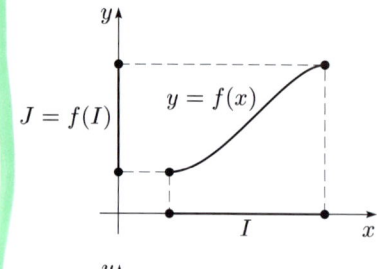

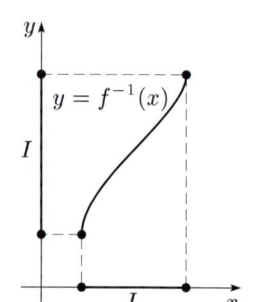

Remarks

1.   The interval I can be *any* type of interval: open or closed, half-open, bounded or unbounded.

2.   There is another version of the Inverse Function Rule with 'strictly increasing' replaced by 'strictly decreasing'.

Example 4.1   Prove that the function

$$f(x) = x^5 + x - 1 \quad (x \in \mathbb{R})$$

has a continuous inverse function, with domain $\mathbb{R}$.

Solution   The domain of f is $\mathbb{R}$, which is an interval.

We know that f is strictly increasing and continuous on $\mathbb{R}$, so conditions 1 and 2 of the Inverse Function Rule hold. Thus the image $J = f(\mathbb{R})$ is an interval, and f has a continuous inverse function

$$f^{-1} \colon J \longrightarrow \mathbb{R},$$

which is strictly increasing on J.

It remains to check that the image J is the whole of $\mathbb{R}$. To prove this, note that $J = f(\mathbb{R})$ contains each of the numbers

$$f(n) = n^5 + n - 1, \quad n \in \mathbb{Z}.$$

Also,

$$f(n) = n^5 + n - 1 \to \infty \text{ as } n \to \infty$$

and

$$f(-n) = -n^5 - n - 1 \to -\infty \text{ as } n \to \infty.$$

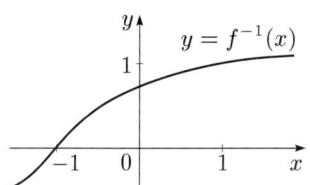

Since J contains each of the intervals $[f(-n), f(n)]$, we deduce that $J = f(\mathbb{R})$ must be $(-\infty, \infty)$; that is, $J = \mathbb{R}$.

Thus f has a continuous inverse function

$$f^{-1} \colon \mathbb{R} \longrightarrow \mathbb{R}. \quad \blacksquare$$

As the above example shows, when we apply the Inverse Function Rule, we have to determine the image $J = f(I)$. Since J is an interval, it is sufficient to determine the *endpoints* of J, which may be real numbers or one of the symbols ∞ and $-\infty$. We must also determine whether or not these endpoints *belong* to J.

For example,
  $(0, 1]$ has endpoints 0 and 1,
and
  $[1, \infty)$ has endpoints 1 and ∞.

Do not let this use of the *symbol* ∞ tempt you to think that ∞ is a real number.

The following diagram illustrates various cases.

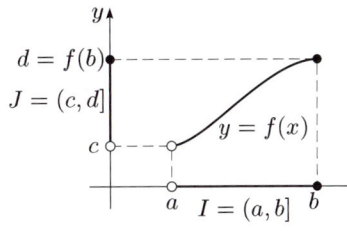 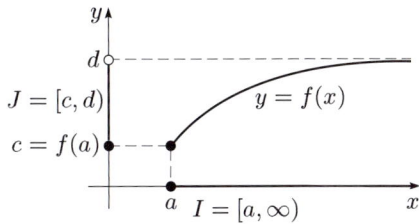

If a is an endpoint of I and $a \in I$, then $c = f(a)$ is the corresponding endpoint of J and $c \in J$.

On the other hand, if a is an endpoint of I and $a \notin I$ (this includes the possibility that a may be ∞ or $-\infty$), then it is a little harder to find the corresponding endpoint c of J. In this case, we use the fact that if $\{a_n\}$ is a monotonic sequence in I and $a_n \to a$, then $f(a_n) \to c$.

Example 4.2   Prove that the function

$$f(x) = x^4 + 2x + 3 \quad (x \in [0, \infty))$$

has a continuous inverse function, with domain $[3, \infty)$.

Solution   The domain of f is $[0, \infty)$, which is an interval.

We know that f is strictly increasing and continuous on $[0, \infty)$, so conditions 1 and 2 of the Inverse Function Rule hold. Thus the image $J = f([0, \infty))$ is an interval, and f has a continuous inverse function

$$f^{-1} \colon J \longrightarrow [0, \infty),$$

which is strictly increasing on J.

See Exercise 1.3(a).

It remains to check that the image J is $[3, \infty)$. For the endpoint 0 of I, we have $0 \in I$, so the corresponding endpoint of J is $f(0) = 3$, and $3 \in J$.

The other endpoint of I is ∞, so to find the corresponding endpoint of J we choose the monotonic sequence $\{n\}$, which lies in I and tends to infinity. We observe that

$$f(n) = n^4 + 2n + 3 \to \infty \quad \text{as } n \to \infty,$$

so the corresponding endpoint of J is ∞. Thus $J = [3, \infty)$, as required. $\blacksquare$

We now summarise the strategy for establishing that a given strictly increasing continuous function f has a continuous inverse function.

Strategy 4.1 To prove that $f \colon I \longrightarrow J$, where I is an interval with endpoints a and b, has a continuous inverse $f^{-1} \colon J \longrightarrow I$.

1. Show that f is strictly increasing on I.

2. Show that f is continuous on I.

3. Determine the endpoint c of J corresponding to the endpoint a of I as follows:

    if $a \in I$, then $f(a) = c$ and $c \in J$,

    if $a \notin I$, then $f(a_n) \to c$ and $c \notin J$,

    where $\{a_n\}$ is a monotonic sequence in I such that $a_n \to a$.

    Determine the endpoint d of J, corresponding to the endpoint b of I, similarly.

There is a corresponding version of Strategy 4.1 if f is strictly decreasing.

In the strictly increasing version the left endpoint of I corresponds to the left endpoint of J, whereas in the strictly decreasing version the left endpoint of I corresponds to the right endpoint of J.

Exercise 4.1 Prove that the function

$$f(x) = x^2 - \frac{1}{x} \quad (x \in (0, \infty))$$

has a continuous inverse function with domain $\mathbb{R}$.

Hint: Use the solution to Exercise 1.3(b).

4.2 Inverses of standard functions

We now use the Inverse Function Rule to define continuous inverse functions for various standard functions. You are already familiar with these inverse functions, but we can now *prove* that they exist and are continuous. For each function, we give brief remarks on the three steps of Strategy 4.1. We also revise some of the properties of these inverse functions.

Here we often use the fact that the restriction of a continuous function is continuous, mentioned on page 21.

nth root function

We asserted the existence of the nth root function at the start of Analysis Block A. We can use Strategy 4.1 to provide a proof of that result.

See Unit AA1, Subsection 5.2.

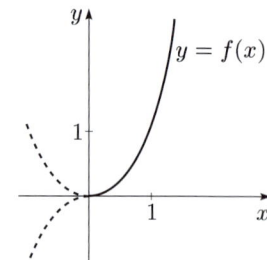

nth root function For any positive integer $n \geq 2$, the function

$$f(x) = x^n \quad (x \in [0, \infty))$$

has a strictly increasing continuous inverse function $f^{-1}(x) = \sqrt[n]{x}$, with domain $[0, \infty)$ and image $[0, \infty)$, called the **nth root function**.

1. f is strictly increasing on $[0, \infty)$.

2. f is continuous on $[0, \infty)$.

3. $f(0) = 0$, and $f(k) = k^n \to \infty$ as $k \to \infty$, so

    $$f([0, \infty)) = [0, \infty).$$

    (We use $\{k\}$ here, to avoid using n for two different purposes in the same expression.)

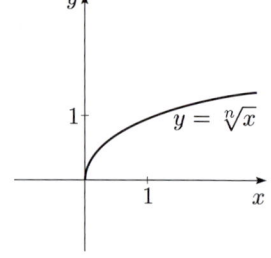

Hence f has a strictly increasing continuous inverse function

$$f^{-1} \colon [0, \infty) \longrightarrow [0, \infty).$$

Inverse trigonometric functions

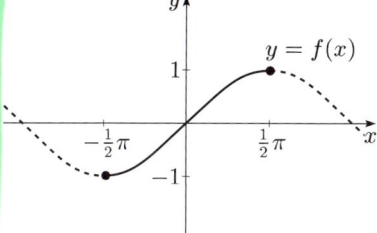

> **sin⁻¹**   The function
>
> $$f(x) = \sin x \quad (x \in [-\tfrac{1}{2}\pi, \tfrac{1}{2}\pi])$$
>
> has a strictly increasing continuous inverse function, with domain $[-1, 1]$ and image $[-\tfrac{1}{2}\pi, \tfrac{1}{2}\pi]$, called **sin⁻¹**.

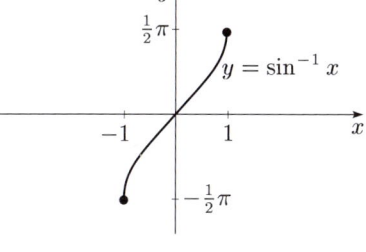

1. The geometric definition of $f(x) = \sin x$ shows that f is strictly increasing on $[-\tfrac{1}{2}\pi, \tfrac{1}{2}\pi]$.
2. f is continuous on $[-\tfrac{1}{2}\pi, \tfrac{1}{2}\pi]$.
3. $\sin(-\tfrac{1}{2}\pi) = -1$ and $\sin(\tfrac{1}{2}\pi) = 1$, so

$$f([-\tfrac{1}{2}\pi, \tfrac{1}{2}\pi]) = [-1, 1].$$

Hence f has a strictly increasing continuous inverse function

$$f^{-1} \colon [-1, 1] \longrightarrow [-\tfrac{1}{2}\pi, \tfrac{1}{2}\pi].$$

The decreasing version of Strategy 4.1 can be applied similarly to prove that the cosine function has an inverse, if we restrict its domain suitably.

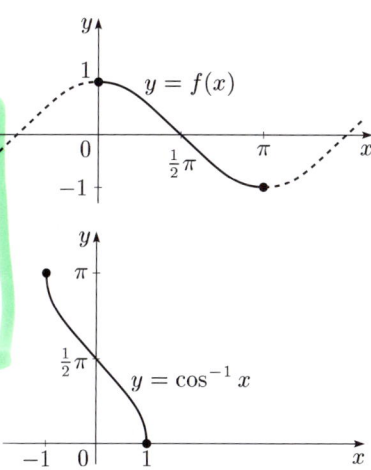

> **cos⁻¹**   The function
>
> $$f(x) = \cos x \quad (x \in [0, \pi])$$
>
> has a strictly decreasing continuous inverse function, with domain $[-1, 1]$ and image $[0, \pi]$, called **cos⁻¹**.

The domain $[0, \pi]$ of f is chosen, by convention, so that f is a restriction of the cosine function which is strictly decreasing and continuous.

Similarly, to form an inverse of the tangent function, we restrict its domain to $(-\tfrac{1}{2}\pi, \tfrac{1}{2}\pi)$, since the tangent function is strictly increasing and continuous on this interval.

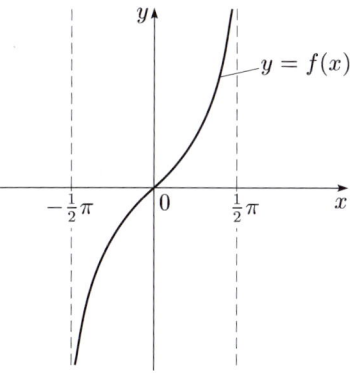

> **tan⁻¹**   The function
>
> $$f(x) = \tan x \quad (x \in (-\tfrac{1}{2}\pi, \tfrac{1}{2}\pi))$$
>
> has a strictly increasing continuous inverse function, with domain $\mathbb{R}$ and image $(-\tfrac{1}{2}\pi, \tfrac{1}{2}\pi)$, called **tan⁻¹**.

In this case, the image set $f((-\tfrac{1}{2}\pi, \tfrac{1}{2}\pi))$ is $\mathbb{R}$ because if $\{a_n\}$ is a monotonic sequence in $(-\tfrac{1}{2}\pi, \tfrac{1}{2}\pi)$ and $a_n \to \tfrac{1}{2}\pi$ as $n \to \infty$, then

$$f(a_n) = \tan a_n = \frac{\sin a_n}{\cos a_n} \to \infty \quad \text{as } n \to \infty.$$

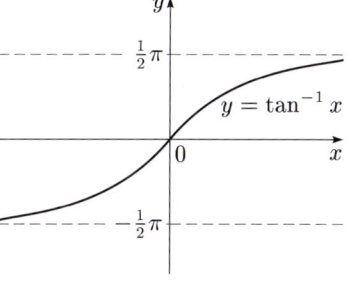

Remark   Some texts use arcsin, arccos and arctan instead of $\sin^{-1}$, $\cos^{-1}$, and $\tan^{-1}$, respectively. These names arise from the geometric definitions of the trigonometric functions. For example, $\arctan x$ is the length of the arc of the circle of radius 1, which corresponds to a tangent line of length x.

Exercise 4.2

  (a) Determine the values of

$$\sin^{-1}(1/\sqrt{2}), \quad \cos^{-1}\left(-\tfrac{1}{2}\right) \quad \text{and} \quad \tan^{-1}(\sqrt{3}).$$

  (b) Prove that

$$\cos(2\sin^{-1}x) = 1 - 2x^2, \quad \text{for } x \in [1,1].$$

  Hint: Let $y = \sin^{-1}x$.

In part (a), use the facts that:
the image of $\sin^{-1}$ is $\left[-\tfrac{1}{2}\pi, \tfrac{1}{2}\pi\right]$,
the image of $\cos^{-1}$ is $[0, \pi]$,
the image of $\tan^{-1}$ is $\left(-\tfrac{1}{2}\pi, \tfrac{1}{2}\pi\right)$.

Inverse function of the exponential function

We now discuss one of the most important inverse functions.

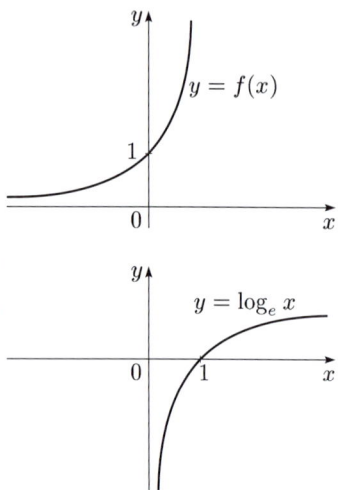

> **$\log_e$**   The function
>
> $$f(x) = e^x \quad (x \in \mathbb{R})$$
>
> has a strictly increasing continuous inverse function f^{-1}, with domain $(0, \infty)$ and image $\mathbb{R}$, called **$\log_e$** or **ln**.

1. f is strictly increasing on $\mathbb{R}$, since

$$x_1 < x_2 \;\Rightarrow\; x_2 - x_1 > 0$$
$$\Rightarrow\; e^{x_2 - x_1} > 1 \quad (\text{since } e^x \geq 1 + x > 1, \text{ for } x > 0)$$
$$\Rightarrow\; e^{x_2} > e^{x_1}.$$

2. f is continuous on $\mathbb{R}$.

3. Since

$$f(n) = e^n \to \infty \;\text{ as } n \to \infty,$$
$$f(-n) = e^{-n} \to 0 \;\text{ as } n \to \infty,$$

the image $f(\mathbb{R}) = (0, \infty)$.

The sequence

$$e^{-n} = (1/e)^n, \quad n = 1, 2, \ldots,$$

is a basic null sequence.

Hence f has a strictly increasing continuous inverse function

$$f^{-1}: (0, \infty) \longrightarrow \mathbb{R}.$$

Exercise 4.3   Prove that

$$\log_e x + \log_e y = \log_e(xy), \quad \text{for } x, y \in (0, \infty).$$

  Hint: Let $a = \log_e x$ and $b = \log_e y$.

Inverse hyperbolic functions

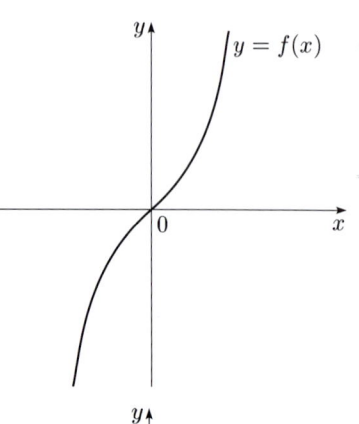

> **$\sinh^{-1}$**   The function
>
> $$f(x) = \sinh x = \tfrac{1}{2}(e^x - e^{-x}) \quad (x \in \mathbb{R})$$
>
> has a strictly increasing continuous inverse function f^{-1}, with domain $\mathbb{R}$ and image $\mathbb{R}$, called **$\sinh^{-1}$**.

1. f is strictly increasing on $\mathbb{R}$, since both the functions

$$x \longmapsto e^x \quad \text{and} \quad x \longmapsto -e^{-x},$$

are strictly increasing on $\mathbb{R}$.

2. f is continuous on $\mathbb{R}$, by the Combination Rules.

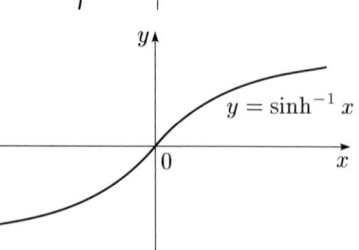

3. Since
$$f(n) = \tfrac{1}{2}(e^n - e^{-n}) \to \infty \quad \text{as } n \to \infty,$$
$$f(-n) = \tfrac{1}{2}(e^{-n} - e^n) \to -\infty \quad \text{as } n \to \infty,$$
the image $f(\mathbb{R}) = \mathbb{R}$.

Hence f has a strictly increasing continuous inverse function
$$f^{-1}\colon \mathbb{R} \longrightarrow \mathbb{R}.$$

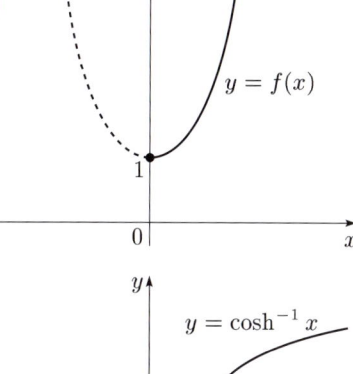

> **cosh^{-1}**   The function
> $$f(x) = \cosh x = \tfrac{1}{2}(e^x + e^{-x}) \quad (x \in [0, \infty))$$
> has a strictly increasing continuous inverse function f^{-1}, with domain $[1, \infty)$ and image $[0, \infty)$, called **cosh^{-1}**.

1. f is strictly increasing on $[0, \infty)$, since $\cosh x = (1 + \sinh^2 x)^{1/2}$ and the function $x \longmapsto \sinh x$ is strictly increasing on $[0, \infty)$.
2. f is continuous on $[0, \infty)$, by the Combination Rules.
3. Since $f(0) = 1$ and
    $$f(n) = \tfrac{1}{2}(e^n + e^{-n}) \to \infty \text{ as } n \to \infty,$$
    the image $f([0, \infty)) = [1, \infty)$.

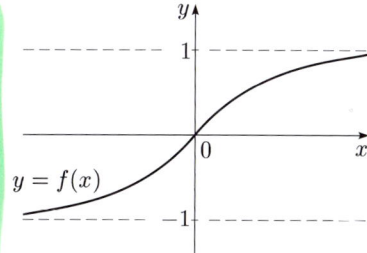

Hence f has a strictly increasing continuous inverse function
$$f^{-1}\colon [1, \infty) \longrightarrow [0, \infty).$$

Strategy 4.1 can be applied in a similar way to show that $f(x) = \tanh x$ is strictly increasing and continuous on $\mathbb{R}$, with $f(\mathbb{R}) = (-1, 1)$.

> **tanh^{-1}**   The function
> $$f(x) = \tanh x = \frac{\sinh x}{\cosh x} \quad (x \in \mathbb{R})$$
> has a strictly increasing continuous inverse function f^{-1}, with domain $(-1, 1)$ and image $\mathbb{R}$, called **tanh^{-1}**.

The inverse hyperbolic functions can all be expressed in terms of $\log_e$, as we show for $\sinh^{-1}$ in the following example.

Example 4.3   Prove that
$$\sinh^{-1} x = \log_e\left(x + \sqrt{x^2 + 1}\right), \quad \text{for } x \in \mathbb{R}.$$

Solution   Let $y = \sinh^{-1} x$, for $x \in \mathbb{R}$, so
$$x = \sinh y = \tfrac{1}{2}(e^y - e^{-y}).$$
Multiplying both sides by e^y, we obtain
$$e^{2y} - 2xe^y - 1 = (e^y)^2 - 2xe^y - 1 = 0.$$
This is a quadratic equation in e^y, with solution
$$e^y = x \pm \sqrt{x^2 + 1}.$$
Since $e^y > 0$, we must choose the $+$ sign. We obtain
$$y = \log_e\left(x + \sqrt{x^2 + 1}\right). \quad \blacksquare$$

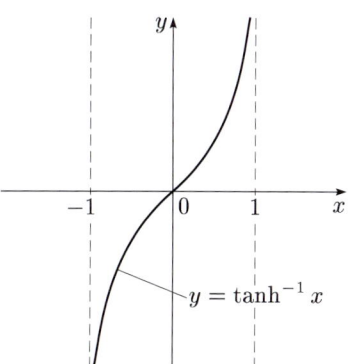

Exercise 4.4   Prove that
$$\cosh^{-1} x = \log_e \left(x + \sqrt{x^2 - 1} \right), \quad \text{for } x \in [1, \infty).$$

4.3 Defining exponential functions

At the start of the course, we asked the following question about the graph $y = 2^x$.

See Unit I1, Subsection 1.2, Frame 25.

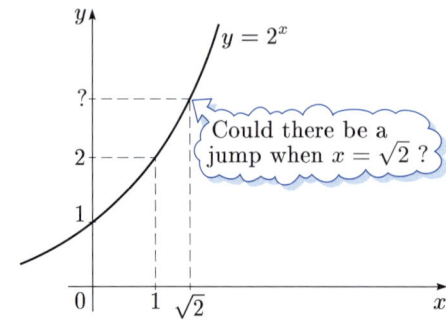

At the start of Analysis Block A, we defined the irrational number $\sqrt{2} = 1.4142\ldots$. We also defined the expression a^x for $a > 0$ when x is a rational number, but *not* when x is irrational. We now provide this missing definition, and also prove that the resulting function $x \longmapsto a^x$ is continuous. In particular, it follows that the graph $y = 2^x$ cannot have any gaps.

See Unit AA1, Section 5.

Recall that
$$e^x = \lim_{n \to \infty} (1 + x/n)^n = \sum_{n=0}^{\infty} \frac{x^n}{n!} \quad (x \geq 0)$$

See Unit AA3, Section 4.

and
$$e^x = \left(e^{-x} \right)^{-1} \quad (x < 0).$$

As we saw in Subsection 4.2, the function $x \longmapsto e^x$ is strictly increasing and continuous, and has a strictly increasing continuous inverse function
$$x \longmapsto \log_e x \quad (x \in (0, \infty)).$$

The function $\log_e$ has the property that

See Exercise 4.3.

$$\log_e(ab) = \log_e a + \log_e b, \quad \text{for } a, b \in (0, \infty).$$

Thus, if $a > 0$ and $n \in \mathbb{N}$, then
$$\log_e(a^n) = n \log_e a, \quad \text{so} \quad a^n = e^{n \log_e a}.$$

With a little further manipulation, we can show that this equation for a^n remains true if n is replaced by any rational number x. Thus it makes sense to *define* a^x, for $a > 0$ and x irrational, by using this equation.

Definition   If $a > 0$, then
$$a^x = e^{x \log_e a} \quad (x \in \mathbb{R}).$$

For example,
$$2^\pi = e^{\pi \log_e 2}.$$

With this definition of a^x, we can verify that the function $x \longmapsto a^x$ is continuous. This follows immediately from the continuity of the function $x \longmapsto e^x$ and the Composition Rule. Moreover, we can also deduce the usual Exponent Laws for a^x from those for e^x. We state these below without proof.

See Unit AA1, Section 5.

Theorem 4.1

(a) If $a > 0$, then the function

$$x \longmapsto a^x = e^{x \log_e a} \quad (x \in \mathbb{R})$$

is continuous.

(b) If $a, b > 0$ and $x, y \in \mathbb{R}$, then

$$a^x b^x = (ab)^x, \quad a^x a^y = a^{x+y} \quad \text{and} \quad (a^x)^y = a^{xy}.$$

In particular, it follows from Theorem 4.1(b) that manipulations such as

$$\left(\sqrt{2}^{\sqrt{2}} \right)^{\sqrt{2}} = \sqrt{2}^{\sqrt{2} \times \sqrt{2}} = \sqrt{2}^2 = 2 \tag{4.1}$$

are justified.

Remark   Equation (4.1) gives an unexpected proof of the result that

there exist irrational numbers a and b such that a^b is rational.

For if $\sqrt{2}^{\sqrt{2}}$ is rational, then we can take $a = b = \sqrt{2}$, but if $\sqrt{2}^{\sqrt{2}}$ is irrational, then (by equation (4.1)) we can take $a = \sqrt{2}^{\sqrt{2}}$ and $b = \sqrt{2}$.

More complicated arguments show that

$$\sqrt{2}^{\sqrt{2}}$$ is irrational.

Exercise 4.5   Use the definition of a^x to prove that each of the following functions is continuous:

(a) $f(x) = x^\alpha \quad (x \in (0, \infty))$, where α is any fixed real number;

(b) $f(x) = x^x \quad (x \in (0, \infty))$.

4.4 Proof of the Inverse Function Rule

In this section, we prove the Inverse Function Rule and we also justify step 3 of Strategy 4.1 for finding the domain of the inverse function.

If you are short of time, omit these proofs.

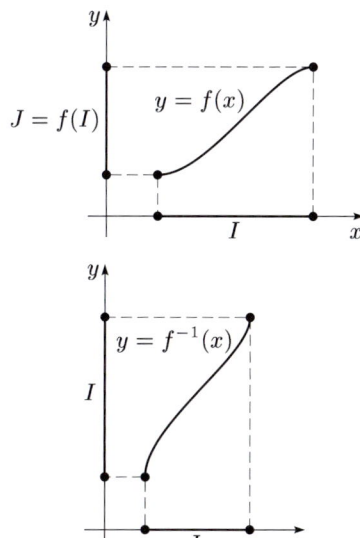

Inverse Function Rule   Let $f : I \longrightarrow J$, where I is an interval and J is the image $f(I)$, be a function such that

1. f is strictly increasing on I;
2. f is continuous on I.

Then J is an interval and f has an inverse function $f^{-1} : J \longrightarrow I$, such that

1'. f^{-1} is strictly increasing on J;
2'. f^{-1} is continuous on J.

Proof   First we prove that $J = f(I)$ is an interval. To prove this, suppose that $y_1, y_2 \in f(I)$ with $y_1 < y_2$ and that y is any number in the interval (y_1, y_2). We want to prove that $y \in f(I)$.

Now $y_1 = f(x_1)$ and $y_2 = f(x_2)$ for some $x_1, x_2 \in I$, with $x_1 < x_2$ because f is strictly increasing. Thus, by the Intermediate Value Theorem, there is a number $x \in (x_1, x_2)$ such that $f(x) = y$. Hence $y \in f(I)$, as required.

Next, the function f is strictly increasing and is therefore one-one. Thus $f^{-1} : J \longrightarrow I$ exists, where $J = f(I)$.

To prove that f^{-1} is strictly increasing on J, we have to show that

$$y_1 < y_2 \Rightarrow f^{-1}(y_1) < f^{-1}(y_2), \quad \text{for } y_1, y_2 \in J.$$

This implication holds because, for $y_1, y_2 \in J$, we have

$$f^{-1}(y_1) \geq f^{-1}(y_2) \Rightarrow f(f^{-1}(y_1)) \geq f(f^{-1}(y_2))$$
$$\Rightarrow y_1 \geq y_2.$$

This is a proof by contraposition; see Unit I2, Subsection 3.7.

Finally, we prove that f^{-1} is continuous on J. Let $y \in J$ and assume, for simplicity, that y is not an endpoint of J. Then $y = f(x)$ for some $x \in I$, and we want to prove that

$$y_n \to y \Rightarrow f^{-1}(y_n) \to f^{-1}(y) = x.$$

Only a slight modification to the argument is needed if y is an endpoint of J.

Thus, we assume that $y_n \to y$ and we want to deduce that

for each $\varepsilon > 0$, there is an integer N such that

$$x - \varepsilon < f^{-1}(y_n) < x + \varepsilon, \quad \text{for all } n > N. \tag{4.2}$$

We know, since f is strictly increasing, that

$$f(x - \varepsilon) < f(x) < f(x + \varepsilon),$$

so, because $y_n \to y = f(x)$, there is an integer N such that

By taking ε small enough, we can assume here that $x - \varepsilon \in I$ and $x + \varepsilon \in I$.

$$f(x - \varepsilon) < y_n < f(x + \varepsilon), \quad \text{for all } n > N.$$

Thus, since f^{-1} is a strictly increasing function, we obtain inequalities (4.2).

This completes the proof of the Inverse Function Rule. ∎

Finally, we justify step 3 of Strategy 4.1 for finding the endpoints of $J = f(I)$.

Strategy 4.1, step 3

Let f satisfy the assumptions of the Inverse Function Rule and let a be an endpoint of I. Determine the corresponding endpoint c of J as follows:

if $a \in I$, then $f(a) = c$ and $c \in J$,

if $a \notin I$, then $f(a_n) \to c$ and $c \notin J$,

where $\{a_n\}$ is a monotonic sequence in I such that $a_n \to a$.

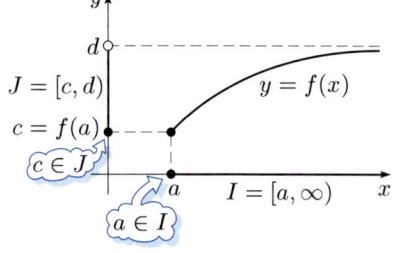

Proof   Suppose that a is the left endpoint of I; the argument for the right endpoint is similar.

If $a \in I$, then $c = f(a) \in J$ and (because f is an increasing function)

$$f(x) \geq f(a) = c, \quad \text{for } x \in I.$$

Thus c is the corresponding left endpoint of J.

If $a \notin I$, then let $\{a_n\}$ be any decreasing sequence in I such that

$$a_n \to a \text{ as } n \to \infty. \tag{4.3}$$

Then $\{f(a_n)\}$ is also a decreasing sequence (because f is an increasing function). Thus, by the Monotonic Sequence Theorem,

See Unit AA2, Theorem 5.2.

$$f(a_n) \to c \text{ as } n \to \infty, \tag{4.4}$$

where c is a real number or $-\infty$.

We now prove that c is an endpoint of J and $c \notin J$.

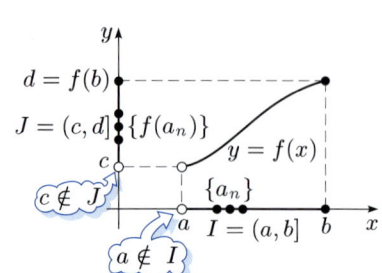

First we show that $(c, f(a_1)) \subset J$. If $c < y < f(a_1)$, then (by equation (4.4)) there exists n such that $f(a_n) < y < f(a_1)$, so $y = f(x)$ for some $x \in (a_n, a_1)$, by the Intermediate Value Theorem, and hence $y \in J$.

Finally, we show that $c \notin J$. Otherwise $c = f(x)$ for some $x > a$, so (by equation (4.3)) there exists n such that $x > a_n$, which implies that $f(x) > f(a_n)$ and hence $f(x) > c$, a contradiction.   ∎

Further exercises

Exercise 4.6   Prove that each of the following functions has a continuous inverse function and determine the domain of the inverse function in each case.

(a) $f(x) = x^3 + 1 - \dfrac{1}{x^2} \quad (x \in (0, \infty))$

(b) $f(x) = \dfrac{1}{(1 + x^3)^2} \quad (x \in [0, \infty))$

Hint: We proved that these functions are monotonic in Exercise 1.6.

Exercise 4.7   State whether or not each of the following statements is true.

(a) $\sin(\sin^{-1} x) = x$, for $x \in [-1, 1]$.

(b) $\sin^{-1}(\sin x) = x$, for $x \in \mathbb{R}$.

Exercise 4.8

(a) Prove that the identity

$$\tan^{-1} x + \tan^{-1} y = \tan^{-1}\left(\frac{x + y}{1 - xy} \right), \quad \text{for } x, y \in \mathbb{R},$$

is true, provided that $\tan^{-1} x + \tan^{-1} y$ lies in $(-\tfrac{1}{2}\pi, \tfrac{1}{2}\pi)$.

(b) Hence evaluate

$$\tan^{-1}(\tfrac{1}{2}) + \tan^{-1}(\tfrac{1}{3}).$$

Solutions to the exercises

1.1 The domain of $f + g$ is $(-\frac{1}{2}\pi, \frac{1}{2}\pi)$; the rule is
$$(f + g)(x) = f(x) + g(x) = e^x + \tan x.$$
The domain of fg is $(-\frac{1}{2}\pi, \frac{1}{2}\pi)$; the rule is
$$(fg)(x) = f(x)g(x) = e^x \tan x.$$
The domain of f/g is $(-\frac{1}{2}\pi, 0) \cup (0, \frac{1}{2}\pi)$; the rule is
$$(f/g)(x) = f(x)/g(x) = e^x / \tan x = e^x \cot x.$$

1.2 The domain of $f \circ g$ is
$$\{x \in \mathbb{R} : \sin x \geq 0\}$$
$$= \cdots \cup [-2\pi, -\pi] \cup [0, \pi] \cup [2\pi, 3\pi] \cup \cdots;$$
the rule is
$$(f \circ g)(x) = \sqrt{\sin x}.$$
The domain of $g \circ f$ is
$$\{x \in [0, \infty) : \sqrt{x} \in \mathbb{R}\} = [0, \infty);$$
the rule is
$$(g \circ f)(x) = \sin \sqrt{x}.$$

1.3 (a) If $0 \leq x_1 < x_2$, then $2x_1 < 2x_2$ and $x_1^4 < x_2^4$. Hence
$$x_1^4 + 2x_1 + 3 < x_2^4 + 2x_2 + 3,$$
so f is strictly increasing, and thus one-one.

(b) If $0 < x_1 < x_2$, then $x_1^2 < x_2^2$ and $1/x_1 > 1/x_2$, so $-1/x_1 < -1/x_2$. Hence
$$x_1^2 - 1/x_1 < x_2^2 - 1/x_2,$$
so f is strictly increasing, and thus one-one.

1.4 (a) The domain of $f + g$ is
$$(-\tfrac{1}{2}\pi, \tfrac{1}{2}\pi) \cap [0, \infty) = [0, \tfrac{1}{2}\pi);$$
the rule is
$$(f + g)(x) = \tan x + \sqrt{x}.$$

(b) As in part (a), the domain of fg is $[0, \frac{1}{2}\pi)$; the rule is
$$(fg)(x) = \sqrt{x} \tan x.$$

(c) The domain of f/g is
$$[0, \tfrac{1}{2}\pi) - \{x : \sqrt{x} = 0\} = (0, \tfrac{1}{2}\pi);$$
the rule is
$$(f/g)(x) = \frac{\tan x}{\sqrt{x}}.$$

(d) The domain of $f \circ g$ is
$$\{x \in [0, \infty) : \sqrt{x} \in (-\tfrac{1}{2}\pi, \tfrac{1}{2}\pi)\} = [0, \tfrac{1}{4}\pi^2);$$
the rule is
$$(f \circ g)(x) = \tan \sqrt{x}.$$

(e) The domain of $g \circ f$ is
$$\{x \in (-\tfrac{1}{2}\pi, \tfrac{1}{2}\pi) : \tan x \in [0, \infty)\} = [0, \tfrac{1}{2}\pi);$$
the rule is
$$(g \circ f)(x) = \sqrt{\tan x}.$$

1.5 First we solve the equation
$$y = \frac{x - 1}{x + 2}$$
to obtain x in terms of y. We find that
$$y = \frac{x - 1}{x + 2} = 1 - \frac{3}{x + 2} \Leftrightarrow x = -2 + \frac{3}{1 - y}.$$
Thus f is one-one, so f has an inverse function with rule $f^{-1}(y) = -2 + 3/(1 - y)$.
For each $x \in (-2, \infty)$, we have $3/(x + 2) > 0$, so $y < 1$. Hence $f((-2, \infty)) \subseteq (-\infty, 1)$.
Also, for each $y \in (-\infty, 1)$, we have $1 - y > 0$, so
$$-2 + \frac{3}{1 - y} \in (-2, \infty).$$
Thus $f((-2, \infty)) \supseteq (-\infty, 1)$, so
$$f((-2, \infty)) = (-\infty, 1).$$
Hence the domain of f^{-1} is $(-\infty, 1)$, so
$$f^{-1}(y) = -2 + \frac{3}{1 - y} \quad (y \in (-\infty, 1)).$$

1.6 (a) If $0 < x_1 < x_2$, then $x_1^3 < x_2^3$ and $-1/x_1^2 < -1/x_2^2$.
Hence
$$x_1^3 + 1 - 1/x_1^2 < x_2^3 + 1 - 1/x_2^2,$$
so f is strictly increasing on $(0, \infty)$.

(b) If $0 \leq x_1 < x_2$, then $0 \leq x_1^3 < x_2^3$, so
$$1 + x_1^3 < 1 + x_2^3.$$
Hence
$$\frac{1}{(1 + x_1^3)^2} > \frac{1}{(1 + x_2^3)^2},$$
so f is strictly decreasing on $[0, \infty)$.

2.1 (a) $\lim_{n \to \infty} 3x_n = 6$, by the Multiple Rule for sequences.

(b) $\lim_{n \to \infty} x_n^2 = 4$, by the Product Rule for sequences.

(c) $\lim_{n \to \infty} 1/x_n = 1/2$, by the Reciprocal Rule for sequences.

2.2 (a) (i) We guess that f is continuous at $a = 2$, so we want to prove that
$$x_n \to 2 \Rightarrow f(x_n) \to f(2). \tag{$*$}$$
Now $f(x_n) = x_n^3 - 2x_n^2$ and $f(2) = 0$. Also,
$$x_n \to 2 \Rightarrow x_n^3 - 2x_n^2 \to 8 - 8 = 0,$$
by the Combination Rules for sequences, so statement $(*)$ holds.

Hence f is continuous at $a = 2$.

(ii) We guess that $f(x) = [x]$ is discontinuous at $a = 1$, since
$$f(x) = 0, \quad \text{for } 0 \le x < 1,$$
and $f(1) = 1$. According to Strategy 2.1 (in Frame 7), we have to find *one* sequence $\{x_n\}$ such that
$$x_n \to 1 \quad \text{but} \quad f(x_n) \nrightarrow f(1).$$
We choose
$$x_n = 1 - \frac{1}{n}, \quad n = 1, 2, \ldots.$$
Then
$$f(x_n) = 0, \quad \text{for } n = 1, 2, \ldots,$$
so we have
$$x_n \to 1 \quad \text{but} \quad f(x_n) \nrightarrow f(1).$$
Hence f is discontinuous at $a = 1$.

(b) (i) To prove that $f(x) = 1$ is continuous at each $a \in \mathbb{R}$, we want to prove that
$$x_n \to a \implies f(x_n) \to f(a). \qquad (*)$$
Now
$$f(x_n) = 1, \quad \text{for } n = 1, 2, \ldots,$$
and
$$f(a) = 1,$$
so statement $(*)$ holds. Hence f is continuous.

(ii) To prove that $f(x) = x$ is continuous at each $a \in \mathbb{R}$, we want to prove that
$$x_n \to a \implies f(x_n) \to f(a). \qquad (*)$$
Now
$$f(x_n) = x_n, \quad \text{for } n = 1, 2, \ldots,$$
and
$$f(a) = a,$$
so statement $(*)$ holds. Hence f is continuous.

2.3 (a) Let $g(x) = \cos x$ and $h(x) = x^2$. Then both g and h are continuous functions and $f = g \circ h$. Hence f is continuous, by the Composition Rule.

(b) To prove that
$$f(x) = \begin{cases} x \sin(1/x), & x \ne 0, \\ 0, & x = 0, \end{cases}$$
is continuous at 0, we use the Squeeze Rule. Since
$$-1 \le \sin(1/x) \le 1, \quad \text{for } x \ne 0,$$
we have
$$-x \le x \sin(1/x) \le x, \quad \text{for } x > 0,$$
and
$$-x \ge x \sin(1/x) \ge x, \quad \text{for } x < 0.$$
Hence
$$-|x| \le x \sin(1/x) \le |x|, \quad \text{for } x \ne 0.$$
Since $f(0) = 0$, we deduce that
$$-|x| \le f(x) \le |x|, \quad \text{for } x \in \mathbb{R}.$$
Thus, if we take $I = \mathbb{R}$, and
$$g(x) = -|x| \quad \text{and} \quad h(x) = |x|,$$
then condition 1 of the Squeeze Rule holds with $a = 0$.

Condition 2 holds, since $g(0) = f(0) = h(0) = 0$.
Condition 3 holds, since g and h are both continuous at 0, by the result of Frame 12 and the Multiple Rule.
Hence f is continuous at 0, by the Squeeze Rule.

2.4 (a) Let $g(x) = e^x$ and $h(x) = -x^2$. Then g and h are both basic continuous functions, so
$$f(x) = 7g(h(x)) = 7e^{-x^2} \quad \text{is continuous,}$$
by the Composition Rule and the Multiple Rule.

(b) Let $g(x) = \sin x$, $h(x) = \sqrt{x}$ and $k(x) = x^2 + 1$. Then g, h and k are basic continuous functions, so
$$g(h(k(x))) = \sin\left(\sqrt{x^2+1}\right) \quad \text{is continuous,}$$
by the Composition Rule (applied twice). Hence
$$f(x) = k(x) + 3g(h(k(x))) \quad \text{is continuous,}$$
by the Combination Rules.

2.5 (a) We prove that
$$f(x) = \begin{cases} x^2 \cos(1/x^2), & x \ne 0, \\ 0, & x = 0, \end{cases}$$
is continuous at 0, using the Squeeze Rule.
Since
$$-1 \le \cos(1/x^2) \le 1, \quad \text{for } x \ne 0,$$
and $x^2 \ge 0$, we have
$$-x^2 \le x^2 \cos(1/x^2) \le x^2, \quad \text{for } x \ne 0.$$
Since $f(0) = 0$, we deduce that
$$-x^2 \le f(x) \le x^2, \quad \text{for } x \in \mathbb{R}.$$
Thus if we take $I = \mathbb{R}$, and
$$g(x) = -x^2 \quad \text{and} \quad h(x) = x^2,$$
then conditions 1, 2 and 3 of the Squeeze Rule hold with $a = 0$.
Hence f is continuous at 0, by the Squeeze Rule.

(b) We prove that
$$f(x) = \begin{cases} \sin(1/x), & x \ne 0, \\ 0, & x = 0, \end{cases}$$
is discontinuous at 0.
According to Strategy 2.1 (in Frame 7), we have to find *one* sequence $\{x_n\}$ such that
$$x_n \to 0 \quad \text{but} \quad f(x_n) \nrightarrow f(0) = 0.$$
We use the fact that $\sin(2n + \frac{1}{2})\pi = 1$, for $n = 0, 1, 2, \ldots$, and choose
$$x_n = \frac{1}{(2n + \frac{1}{2})\pi}, \quad n = 0, 1, 2, \ldots.$$
Then $x_n \to 0$ and
$$\begin{aligned} f(x_n) &= \sin(1/x_n) \\ &= \sin(2n + \tfrac{1}{2})\pi = 1, \quad \text{for } n = 0, 1, 2, \ldots, \end{aligned}$$
so
$$f(x_n) \nrightarrow f(0) = 0.$$
Hence f is discontinuous at 0.

2.6 Let $I = \mathbb{R}$ and define the basic continuous functions

$$g(x) = e^x \quad \text{and} \quad h(x) = \frac{1}{x+1}.$$

Then f is defined on I and $0 \in I$. Also,

$$f(x) = g(x), \quad \text{for } x \in (-\infty, 0)$$

and

$$f(x) = h(x), \quad \text{for } x \in (0, \infty),$$

so condition 1 of the Glue Rule holds with $a = 0$. Moreover, $f(0) = g(0) = h(0) = 1$, so condition 2 holds, and g, h are continuous at 0, so condition 3 holds.

Hence f is continuous at 0, by the Glue Rule.

2.7 (a) Let $g(x) = x^2 + 1$, $h(x) = \sin x$ and $k(x) = e^x$. Then g, h and k are basic continuous functions. Hence, by the Composition Rule,

$$f = k \circ h \circ g \quad \text{is continuous.}$$

(b) Let $g(x) = \sqrt{x}$, $h(x) = e^x$ and $k(x) = x^5$. Then g, h and k are basic continuous functions. Hence, by the Composition Rule and the Sum Rule,

$$f = h \circ g + k \quad \text{is continuous.}$$

2.8 (a) The function f is *not* continuous at 0. For example, if

$$x_n = -\frac{1}{n}, \quad n = 1, 2, \ldots,$$

then $x_n < 0$, for $n = 1, 2, \ldots$, and $x_n \to 0$, so

$$f(x_n) = x_n \to 0.$$

Since $f(0) = 1/(0+1) = 1$, we have

$$f(x_n) \nrightarrow f(0),$$

which shows that f is discontinuous at 0.

(b) We prove that

$$f(x) = \begin{cases} \sin x \sin(1/x), & x \neq 0, \\ 0, & x = 0, \end{cases}$$

is continuous at 0, by using the Squeeze Rule. Now

$$-1 \leq \sin(1/x) \leq 1, \quad \text{for } x \neq 0,$$

so

$$-\sin x \leq \sin x \sin(1/x) \leq \sin x, \quad \text{for } x \in (0, \pi),$$

since $\sin x \geq 0$, for $x \in (0, \pi)$.

Also,

$$-\sin x \geq \sin x \sin(1/x) \geq \sin x, \quad \text{for } x \in (-\pi, 0),$$

since $\sin x \leq 0$, for $x \in (-\pi, 0)$.

Hence

$$-|\sin x| \leq \sin x \sin(1/x) \leq |\sin x|, \quad \text{for } 0 < |x| < \pi.$$

Since $f(0) = 0$, we deduce that

$$-|\sin x| \leq f(x) \leq |\sin x|, \quad \text{for } x \in (-\pi, \pi),$$

with equality when $x = 0$.

Thus, if we take $I = (-\pi, \pi)$, $a = 0$, and

$$g(x) = -|\sin x| \quad \text{and} \quad h(x) = |\sin x|,$$

then g and h are continuous at 0, by the Composition Rule and Multiple Rule, so conditions 1, 2 and 3 of the Squeeze Rule hold.

Hence f is continuous at 0, by the Squeeze Rule.

(c) The function

$$f(x) = \begin{cases} (1/x)\cos(1/x), & x \neq 0, \\ 0, & x = 0, \end{cases}$$

is not continuous at 0. For example, if

$$x_n = \frac{1}{2n\pi}, \quad n = 0, 1, 2, \ldots,$$

then $x_n \to 0$, but

$$f(x_n) = \frac{1}{x_n} \cos\left(\frac{1}{x_n}\right) = 2n\pi \to \infty,$$

so $f(x_n) \nrightarrow f(0) = 0$, as required.

(d) Let $I = \mathbb{R}$ and define the basic continuous functions

$$g(x) = 0 \quad \text{and} \quad h(x) = \sqrt{x}.$$

Then f is defined on I and $0 \in I$. Also,

$$f(x) = g(x), \quad \text{for } x \in (-\infty, 0),$$
$$f(x) = h(x), \quad \text{for } x \in (0, \infty),$$

so condition 1 of the Glue Rule holds with $a = 0$. Moreover, $f(0) = g(0) = h(0) = 0$, so condition 2 holds, and g, h are continuous at 0, so condition 3 holds.

Hence f is continuous at 0, by the Glue Rule.

2.9 To prove the continuity of f at $a = \frac{1}{2}\pi$, take $I = (-\frac{1}{2}\pi, \infty)$ and define the basic continuous functions

$$g(x) = \sin x \quad \text{and} \quad h(x) = 1.$$

Then f is defined on I and $\frac{1}{2}\pi \in I$. Also,

$$f(x) = g(x), \quad \text{for } x \in (-\tfrac{1}{2}\pi, \tfrac{1}{2}\pi),$$
$$f(x) = h(x), \quad \text{for } x \in (\tfrac{1}{2}\pi, \infty),$$

so condition 1 of the Glue Rule holds with $a = \frac{1}{2}\pi$. Moreover,

$$g(\tfrac{1}{2}\pi) = f(\tfrac{1}{2}\pi) = h(\tfrac{1}{2}\pi) = 1,$$

so condition 2 holds, and g, h are continuous at $\frac{1}{2}\pi$, so condition 3 holds.

We deduce that f is continuous at $\frac{1}{2}\pi$, by the Glue Rule.

A similar argument can be given for $a = -\frac{1}{2}\pi$, with $I = (-\infty, \frac{1}{2}\pi)$, $g(x) = -1$ and $h(x) = \sin x$. (Alternatively, note that f is an odd function:

$$f(x) = -f(-x), \quad \text{for } x \in \mathbb{R},$$

so the continuity of f at $-\frac{1}{2}\pi$ follows from its continuity at $\frac{1}{2}\pi$, by the Composition Rule and Multiple Rule.)

3.1 We know that c lies in $(\frac{1}{2}, 1)$, so we calculate
$$f(\tfrac{3}{4}) = (\tfrac{3}{4})^5 + \tfrac{3}{4} - 1 \simeq -0.0127 < 0.$$
Thus c lies in $(\frac{3}{4}, 1)$, so we calculate
$$f(\tfrac{7}{8}) = (\tfrac{7}{8})^5 + \tfrac{7}{8} - 1 \simeq 0.388 > 0.$$
Thus c lies in $(\frac{3}{4}, \frac{7}{8})$, so we calculate
$$f(\tfrac{13}{16}) = (\tfrac{13}{16})^5 + \tfrac{13}{16} - 1 \simeq 0.167 > 0.$$
Thus c lies in $(\frac{3}{4}, \frac{13}{16})$, an interval of length $\frac{1}{16}$.

3.2 We consider the function
$$f(x) = \cos x - x$$
and show that f has a zero c in $(0, 1)$. Now f is continuous, by the Combination Rules, and
$$f(0) = \cos 0 - 0 = 1 > 0$$
and
$$f(1) = \cos 1 - 1 < 0.$$
Thus, by the Intermediate Value Theorem, there is a number c in $(0, 1)$ such that
$$f(c) = 0, \quad \text{so} \quad \cos c = c.$$

3.3 If $f(0) = 0$ or $f(1) = 1$, then we can take $c = 0$ or $c = 1$, respectively.

Otherwise, we have $f(0) > 0$ and $f(1) < 1$, since $0 \le f(x) \le 1$, for $0 \le x \le 1$.

Now we consider the function
$$g(x) = f(x) - x \quad (x \in [0, 1])$$
and show that g has a zero c in $(0, 1)$.

Then g is continuous on $[0, 1]$, by the Combination Rules, and
$$g(0) = f(0) - 0 > 0$$
and
$$g(1) = f(1) - 1 < 0.$$
Thus, by the Intermediate Value Theorem, there is a number c in $(0, 1)$ such that
$$g(c) = 0, \quad \text{so} \quad f(c) = c.$$

3.4 We have
$$p(-1) = -3, \quad p(0) = 1, \quad p(1) = -1, \quad p(2) = 3,$$
so
$$p(-1) < 0 < p(0),$$
$$p(0) > 0 > p(1),$$
$$p(1) < 0 < p(2).$$
Since p is continuous, we deduce by the Intermediate Value Theorem that p has a zero in each of the intervals
$$(-1, 0), \quad (0, 1), \quad (1, 2).$$

3.5 When
$$p(x) = x^5 + 3x^4 - x - 1 \quad (x \in \mathbb{R}),$$
we have
$$M = 1 + \max\{|3|, |-1|, |-1|\} = 4,$$
so all the zeros of p lie in $(-4, 4)$, by Theorem 3.3.

Calculating $p(n)$ for integers n in $[-4, 4]$, we obtain

n	-4	-3	-2	-1	0	1	2
$p(n)$	-253	2	17	2	-1	2	77

Thus p changes sign on each of the intervals
$$[-4, -3], \quad [-1, 0], \quad [0, 1].$$
Since p is continuous, we deduce by the Intermediate Value Theorem that p has a zero in each of the intervals
$$(-4, -3), \quad (-1, 0), \quad (0, 1).$$

3.6 (a) When
$$p(x) = x^4 - 4x^3 + 3x^2 + 2x - 1,$$
we have
$$M = 1 + \max\{|-4|, |3|, |2|, |-1|\} = 5,$$
so all the zeros of p lie in $(-5, 5)$, by Theorem 3.3.

Calculating $p(n)$ for integers n in $[-5, 5]$, we obtain

n	-2	-1	0	1	2	3
$p(n)$	55	5	-1	1	-1	5

Thus p changes sign on each of the intervals
$$[-1, 0], \quad [0, 1], \quad [1, 2], \quad [2, 3].$$
Since p is continuous, we deduce by the Intermediate Value Theorem that p has a zero in each of the intervals
$$(-1, 0), \quad (0, 1), \quad (1, 2), \quad (2, 3).$$

(b) When
$$p(x) = 3x^3 - 8x^2 + x + 3$$
$$= 3\left(x^3 - \tfrac{8}{3}x^2 + \tfrac{1}{3}x + 1\right),$$
we have
$$M = 1 + \max\{|-\tfrac{8}{3}|, |\tfrac{1}{3}|, |1|\} = \tfrac{11}{3},$$
so all the zeros of p lie in $(-\frac{11}{3}, \frac{11}{3})$, by Theorem 3.3.

Calculating $p(n)$ for integers n in $[-4, 4]$, we obtain

n	-2	-1	0	1	2	3
$p(n)$	-55	-9	3	-1	-3	15

Thus p changes sign on each of the intervals
$$[-1, 0], \quad [0, 1], \quad [2, 3].$$
Since p is continuous, we deduce by the Intermediate Value Theorem that p has a zero in each of the intervals
$$(-1, 0), \quad (0, 1), \quad (2, 3).$$

3.7 By calculation,
$$f(\tfrac{2}{3}\pi) = \tfrac{2}{3}\pi - \sin(\tfrac{2}{3}\pi) - \tfrac{2}{3}\pi$$
$$= -\sin(\tfrac{1}{3}\pi) = -\sqrt{3}/2 < 0$$
and
$$f(\tfrac{5}{6}\pi) = \tfrac{5}{6}\pi - \sin(\tfrac{5}{6}\pi) - \tfrac{2}{3}\pi$$
$$= \tfrac{1}{6}\pi - \sin(\tfrac{1}{6}\pi) = \tfrac{1}{6}\pi - \tfrac{1}{2} > 0,$$
because $\pi > 3$.

Also, f is continuous by the Combination Rules.

Hence, by the Intermediate Value Theorem, f has a zero in $(\frac{2}{3}\pi, \frac{5}{6}\pi)$.

3.8 If $f(0) = 0$ or $f(1) = 1$, then we can take $c = 0$ or $c = 1$, respectively.

Otherwise, we have $f(0) > 0$ and $f(1) < 1$, since $0 \le f(x) \le 1$, for $0 \le x \le 1$.

Now we consider the function
$$g(x) = f(x) - x^3 \quad (x \in [0,1]).$$
Then g is continuous on $[0,1]$, by the Combination Rules, and
$$g(0) = f(0) - 0 > 0$$
and
$$g(1) = f(1) - 1 < 0.$$
Thus, by the Intermediate Value Theorem, there is a number c in $(0,1)$ such that
$$g(c) = 0, \quad \text{so} \quad f(c) = c^3.$$

3.9 (a) Let $|x| > 1$, so $1/|x| < 1$. Using the Triangle Inequality, we obtain
$$|r(x)| = \left| \frac{a_{n-1}}{x} + \cdots + \frac{a_1}{x^{n-1}} + \frac{a_0}{x^n} \right|$$
$$\le \left| \frac{a_{n-1}}{x} \right| + \cdots + \left| \frac{a_1}{x^{n-1}} \right| + \left| \frac{a_0}{x^n} \right|$$
$$\le K \left(\frac{1}{|x|} + \cdots + \frac{1}{|x|^{n-1}} + \frac{1}{|x|^n} \right)$$
$$< K \left(\frac{1}{|x|} + \frac{1}{|x|^2} + \cdots \right)$$
$$= \frac{K(1/|x|)}{1 - 1/|x|} = \frac{K}{|x| - 1},$$
by summing the convergent geometric series. Hence
$$|r(x)| < \frac{K}{|x| - 1}, \quad \text{for } |x| > 1.$$
Thus, if
$$|x| \ge M = 1 + \max\{|a_{n-1}|, \ldots, |a_1|, |a_0|\} = 1 + K,$$
then $|x| - 1 > K$, so $|r(x)| < 1$, as required.

(b) We have
$$p(x) = x^n(1 + r(x)), \quad \text{for } x \ne 0.$$
By part (a),
$$1 + r(x) > 0, \quad \text{for } |x| \ge M.$$
Thus $p(x)$ has the same sign as x^n, for $|x| \ge M$. Hence any zero of p must lie in $(-M, M)$, as required.

(c) If n is odd, then x^n is positive for $x \ge M$ and negative for $x \le -M$. Hence, by part (b), $p(x)$ must be positive for $x \ge M$ and negative for $x \le -M$. By the Intermediate Value Theorem, therefore, p must have a zero in $(-M, M)$.

(d) Using the hint, we have
$$q(x) = x^n + a_{n-1}x^{n-1} + \cdots + a_1 x.$$
Thus, since n is even, it follows from part (b) that
$$q(x) > 0, \quad \text{for } |x| \ge M. \tag{S.1}$$

Now q is continuous on the closed interval $[-M, M]$ so, by the Extreme Value Theorem, there exists $c \in [-M, M]$ such that
$$q(x) \ge q(c), \quad \text{for } x \in [-M, M]. \tag{S.2}$$
In particular, we have $q(c) \le q(0) = 0$.
Thus, by equations (S.1) and (S.2),
$$q(x) \ge q(c), \quad \text{for } x \in \mathbb{R},$$
so
$$p(x) = q(x) + a_0 \ge q(c) + a_0 = p(c), \quad \text{for } x \in \mathbb{R},$$
as required.

4.1 We use Strategy 4.1.
1. We showed that f is strictly increasing on $(0, \infty)$ in Exercise 1.3(b).
2. The function
$$f(x) = x^2 - \frac{1}{x} = \frac{x^3 - 1}{x} \quad (x \in (0, \infty))$$
is the restriction to $(0, \infty)$ of a rational function which is continuous on $\mathbb{R} - \{0\}$. Hence f is continuous.
3. Now choose the increasing sequence $\{n\}$, which tends to ∞, the right endpoint of $(0, \infty)$. Then
$$f(n) = n^2 - 1/n \to \infty \quad \text{as } n \to \infty,$$
by the Reciprocal Rule; see Unit AA2, Section 4. Thus the right endpoint of $J = f((0, \infty))$ is ∞.
Then choose the decreasing sequence $\{1/n\}$, which tends to 0, the left endpoint of $(0, \infty)$. Then
$$f(1/n) = 1/n^2 - n \to -\infty \quad \text{as } n \to \infty,$$
by the Reciprocal Rule. Thus the left endpoint of $J = f((0, \infty))$ is $-\infty$.
Hence $J = (-\infty, \infty) = \mathbb{R}$, so f has a continuous inverse function
$$f^{-1} \colon \mathbb{R} \longrightarrow (0, \infty),$$
by the Inverse Function Rule.

4.2 (a) Since $\sin(\frac{1}{4}\pi) = 1/\sqrt{2}$ and $\frac{1}{4}\pi$ lies in $[-\frac{1}{2}\pi, \frac{1}{2}\pi]$, we have
$$\sin^{-1}(1/\sqrt{2}) = \tfrac{1}{4}\pi.$$
Since $\cos(\frac{2}{3}\pi) = -\frac{1}{2}$ and $\frac{2}{3}\pi$ lies in $[0, \pi]$,
$$\cos^{-1}(-\tfrac{1}{2}) = \tfrac{2}{3}\pi.$$
Since $\tan(\frac{1}{3}\pi) = \sqrt{3}$ and $\frac{1}{3}\pi$ lies in $(-\frac{1}{2}\pi, \frac{1}{2}\pi)$,
$$\tan^{-1}(\sqrt{3}) = \tfrac{1}{3}\pi.$$

(b) Following the hint, we put $y = \sin^{-1} x$. Then
$$\cos(2\sin^{-1} x) = \cos(2y) = 1 - 2\sin^2 y = 1 - 2x^2,$$
since $x = \sin y$.

4.3 Following the hint, we put $a = \log_e x$ and $b = \log_e y$. Then $x = e^a$ and $y = e^b$, so
$$\log_e(xy) = \log_e(e^a e^b)$$
$$= \log_e(e^{a+b}) = a + b = \log_e x + \log_e y.$$

4.4 Let $y = \cosh^{-1} x$, where $x \geq 1$. Then
$$x = \cosh y = \tfrac{1}{2}(e^y + e^{-y}).$$
Hence
$$e^{2y} - 2xe^y + 1 = (e^y)^2 - 2xe^y + 1 = 0.$$
This is a quadratic equation in e^y, with solutions
$$e^y = x \pm \sqrt{x^2 - 1}.$$
Both choices of $\pm$ give a positive expression on the right, but we also have $e^y \geq 1$, since $y \geq 0$.
Since
$$\left(x + \sqrt{x^2 - 1}\right)\left(x - \sqrt{x^2 - 1}\right) = x^2 - (x^2 - 1) = 1$$
and $x + \sqrt{x^2 - 1} \geq 1$, we have $x - \sqrt{x^2 - 1} \leq 1$.
Thus we choose the $+$ sign, to give
$$y = \log_e \left(x + \sqrt{x^2 - 1}\right).$$
(The value $y = \log_e \left(x - \sqrt{x^2 - 1}\right)$ gives the negative solution of the equation $\cosh y = x$.)

4.5 (a) For $x > 0$, we have
$$f(x) = x^\alpha = e^{\alpha \log_e x}.$$
Now the functions
$$x \longmapsto \log_e x \quad (x \in (0, \infty)),$$
$$x \longmapsto e^x \quad (x \in \mathbb{R}),$$
are both continuous, so f is continuous by the Multiple Rule and the Composition Rule.

(b) For $x > 0$, we have
$$f(x) = x^x = e^{x \log_e x}.$$
Now the functions
$$x \longmapsto \log_e x \quad (x \in (0, \infty)),$$
$$x \longmapsto x \quad (x \in \mathbb{R}),$$
$$x \longmapsto e^x \quad (x \in \mathbb{R}),$$
are all continuous, so f is continuous by the Product Rule and the Composition Rule.

4.6 (a) We use Strategy 4.1.
1. We showed that f is strictly increasing on $(0, \infty)$ in Exercise 1.6(a).
2. The function
$$f(x) = x^3 + 1 - \frac{1}{x^2} = \frac{x^5 + x^2 - 1}{x^2} \quad (x \in (0, \infty))$$
is the restriction to $(0, \infty)$ of a rational function and is therefore continuous.
3. Choose the increasing sequence $\{n\}$, which tends to ∞. Then
$$f(n) = n^3 + 1 - \frac{1}{n^2} \to \infty \text{ as } n \to \infty,$$
by the Reciprocal Rule; see Unit AA2, Section 4. Thus the right endpoint of $J = f((0, \infty))$ is ∞.
Choose the decreasing sequence $\{1/n\}$, which tends to 0. Then
$$f(1/n) = 1/n^3 + 1 - n^2 \to -\infty \text{ as } n \to \infty,$$
by the Reciprocal Rule. Thus the left endpoint of J is $-\infty$.

Hence $J = (-\infty, \infty) = \mathbb{R}$, so f has a continuous inverse function
$$f^{-1} \colon \mathbb{R} \longrightarrow (0, \infty),$$
by the Inverse Function Rule.

(b) We use Strategy 4.1.
1. We showed that f is strictly decreasing on $[0, \infty)$ in Exercise 1.6(b).
2. The function
$$f(x) = \frac{1}{(1 + x^3)^2} \quad (x \in [0, \infty))$$
is the restriction to $[0, \infty)$ of a rational function and is therefore continuous.
3. We have $f(0) = 1$, so the corresponding endpoint of $J = f([0, \infty))$ is 1, and $1 \in J$.
Choose the increasing sequence $\{n\}$, which tends to ∞. Then
$$f(n) = \frac{1}{(1 + n^3)^2} \to 0 \text{ as } n \to \infty.$$
Thus the corresponding endpoint of J is 0, and $0 \notin J$.
Hence $J = (0, 1]$, so f has a continuous inverse function
$$f^{-1} \colon (0, 1] \longrightarrow [0, \infty),$$
by the Inverse Function Rule.

4.7 (a) True: by definition,
$$\sin(\sin^{-1} x) = x, \quad \text{for } x \in [-1, 1].$$

(b) False: for example,
$$\sin^{-1}(\sin 2\pi) = \sin^{-1} 0 = 0.$$

4.8 (a) Let $a = \tan^{-1} x$ and $b = \tan^{-1} y$. Then $x = \tan a$ and $y = \tan b$, so
$$\tan(a + b) = \frac{\tan a + \tan b}{1 - \tan a \tan b} = \frac{x + y}{1 - xy}.$$
Hence
$$a + b = \tan^{-1}\left(\frac{x + y}{1 - xy}\right),$$
provided that $a + b$ lies in $(-\tfrac{1}{2}\pi, \tfrac{1}{2}\pi)$, which is the image of $\tan^{-1}$. Thus
$$\tan^{-1} x + \tan^{-1} y = \tan^{-1}\left(\frac{x + y}{1 - xy}\right),$$
provided that $\tan^{-1} x + \tan^{-1} y$ lies in $(-\tfrac{1}{2}\pi, \tfrac{1}{2}\pi)$.

(b) Since $\tan^{-1}$ is strictly increasing, we have
$$0 < \tan^{-1}(\tfrac{1}{2}) + \tan^{-1}(\tfrac{1}{3}) < 2\tan^{-1}(1) = \tfrac{1}{2}\pi.$$
Hence, we can apply the formula in part (a):
$$\tan^{-1}(\tfrac{1}{2}) + \tan^{-1}(\tfrac{1}{3}) = \tan^{-1}\left(\frac{\tfrac{1}{2} + \tfrac{1}{3}}{1 - \tfrac{1}{6}}\right)$$
$$= \tan^{-1}(1) = \tfrac{1}{4}\pi.$$

Index